The Competitive Status of the U.S. Electronics Industry

A Study of the Influences of Technology in Determining International Industrial Competitive Advantage

Prepared by the Electronics Panel,
Committee on Technology and
International Economic and Trade Issues

of the Office of the Foreign Secretary,
National Academy of Engineering

and the Commission on Engineering and
Technical Systems, National
Research Council

John G. Linvill, Chairman
Annette M. LaMond and
Robert W. Wilson, Rapporteurs

NATIONAL ACADEMY PRESS
Washington, D.C. 1984

National Academy Press ● 2101 Constitution Avenue, N.W. ● Washington, D.C. 20418

This project was supported under Master Agreement No. 79-02702 between the National Science Foundation and the National Academy of Sciences.

Library of Congress Catalog Card Number 83-83127
International Standard Book Number 0-309-03397-7

Printed in the United States of America

Participants at Meetings of the Electronics Panel,
Committee on Technology and
International Economic and Trade Issues

Panel

JOHN G. LINVILL (Chairman), Professor, Department of
Electrical Engineering, Stanford University
FERNANDO J. CORBATO, Professor of Computer Science and
Engineering, Massachusetts Institute of Technology
THERESE FLAHERTY, Assistant Professor, Harvard University
Business School
EUGENE I. GORDON, Director, Lightwave Devices Laboratory,
Bell Laboratories
WILLIAM C. HITTINGER, Executive Vice-President, Research and
Engineering, RCA Corporation
JOSEPH C. R. LICKLIDER, Professor, Massachusetts Institute of
Technology
ROBERT N. NOYCE, Vice-Chairman of the Board, INTEL
Corporation
DANIEL I. OKIMOTO, Assistant Professor, Department of
Political Science, Stanford University
M. KENNETH OSHMAN, President and Chief Executive Officer,
ROLM Corporation
MICHAEL RADNOR, Director, Center for the Interdisciplinary
Study of Science and Technology, Northwestern University
WILLIAM V. RAPP, Vice-President, Project Services Group, Bank
of America
ROGER S. SEYMOUR, Consultant*

Rapporteurs

ANNETTE M. LaMOND, Consultant-Economics and Competitive
Strategy, Cambridge, MA

---

*Formerly, Program Director, IBM Corporation

iii

ROBERT W. WILSON, Consultant-Economics and Competitive Strategy, Weston, MA

Additional Participants

JOHN ALIC, Project Director of International Security of Commerical Programs, Office of Technology Assessment, U.S. Congress

CLAUDE BANDY, Science and Electronics Division, U.S. Department of Commerce

ARDEN L. BEMENT, JR., Resource and Advanced Technology, U.S. Department of Defense

NAZIR BHAGAT, Office of Productivity, Technology, and Innovation, U.S. Department of Commerce

JOHN CARR, Japan-United States Economic Relations Group

FRANK CAUFIELD, Kleiner Perkins Caufield & Byers

JACK CLIFFORD, Office of Producer Goods, U.S. Department of Commerce

LIAM FAHEY, Center for the Interdisciplinary Study of Science and Technology, Northwestern University

PETER F. FROST, Office of Advanced Technology, U.S. Department of State

JAMES GANNON, NBC News, New York

JEFFREY A HART, Professional Staff, President's Commission for a National Agenda for the Eighties

PHIL MARCUS, Office of Producer Goods, U.S. Department of Commerce

EGILS MILBERGS, Director of Productivity, Technology, and Innovation, U.S. Department of Commerce

JOHN McPHEE, Office of Producer Goods, U.S. Department of Commerce

SUMIYE OKUBO, Policy Analyst, Division of Policy Research and Analysis, Scientific, Technological, and International Affairs, National Science Foundation

SAUL PADWO, Director of Licensing, Office of Export Administration, U.S. Department of Commerce

WILLIAM J. PERRY, Under Secretary of Defense for Research and Development, Office of the Secretary of Defense

ROLF P. PIEKARZ, Senior Policy Analyst, Division of Policy Research and Analysis, Scientific, Technological, and International Affairs, National Science Foundation

ALAN RAPOPORT, Policy Analyst, Division of Policy Research and Analysis, Scientific, Technological, and International Affairs, National Science Foundation

Committee on Technology and
International Economic and Trade Issues (CTIETI)

Chairman

N. BRUCE HANNAY, National Academy of Engineering Foreign
    Secretary and Vice-President, Research and Patents, Bell
    Laboratories (retired)

Members

WILLIAM J. ABERNATHY, Professor, Harvard University
    Graduate School of Business Administration and Chairman,
    CTIETI Automobile Panel
JACK N. BEHRMAN, Luther Hodges Distinguished Professor of
    International Business, University of North Carolina
CHARLES C. EDWARDS, President, Scripps Clinic and Research
    Foundation and Chairman, CTIETI Pharmaceutical Panel
W. DENNEY FREESTON, JR., Associate Dean, College of
    Engineering, Georgia Institute of Technology and Chairman,
    CTIETI Fibers, Textiles, and Apparel Panel
JERRIER A. HADDAD, Vice-President, Technical Personnel
    Development, IBM Corporation (retired)
MILTON KATZ, Henry L. Stimson Professor of Law Emeritus,
    Harvard Law School
RALPH LANDAU, Chairman, Listowel Incorporated* and
    Vice-President, National Academy of Engineering
JOHN G. LINVILL, Professor, Department of Electrical
    Engineering, Stanford University and Chairman, CTIETI
    Electronics Panel

---

*Formerly, Chairman of the Board, Halcon-SD Group.

vii

RAY McCLURE, Program Leader, Precisions Engineering Program, Lawrence Livermore Laboratory and Chairman, CTIETI Machine Tools Panel

BRUCE S. OLD, President, Bruce S. Old Associates, Inc. and Chairman, CTIETI Ferrous Metals Panel

MARKLEY ROBERTS, Economist, AFL-CIO

LOWELL W. STEELE, Consultant-Technology Planning and Management*

MONTE C. THRODAHL, Vice-President, Technology, Monsanto Company

---

*Formerly, Staff Executive, Corporate Technology Planning, General Electric Company

Preface

In August 1976 the Committee on Technology and International Economic and Trade Issues examined a number of technological issues and their relationship to the potential entrepreneurial vitality of the U.S. economy. The committee was concerned with the following:

- Technology and its effect on trade between the United States and the other countries of the Organization for Economic Cooperation and Development (OECD).
- Relationships between technological innovation and U.S. productivity and competitiveness in world trade; impacts of technology and trade on U.S. levels of employment.
- Effects of technology transfer on the development of the less-developed countries (LDCs) and the impact of this transfer on U.S. trade with these nations.
- Trade and technology exports in relation to U.S. national security.

In its 1978 report, Technology, Trade, and the U.S. Economy,* the committee concluded that the state of the nation's competitive position in world trade is a reflection of the health of the domestic economy. The committee stated that, as a consequence, the improvement of our position in international trade depends primarily upon improvement of the domestic economy. The committee further concluded that one of the major factors affecting the health of our domestic economy is the state of

---

*National Research Council, 1978. Technology, Trade, and the U.S. Economy. Report of a workshop held at Woods Hole, Massachusetts, August 22-31, 1976. National Academy of Sciences, Washington, D.C.

industrial innovation. Considerable evidence was presented during the study to indicate that the innovation process in the United States is not as vigorous as it once was. The committee recommended that further work be undertaken to provide a more detailed examination of the U.S. government policies and practices that may bear on technological innovation.

The first phase of the study based on the original recommendations resulted in a series of published monographs that addressed government policies in the following areas:

- The International Technology Transfer Process.*
- The Impact of Regulation on Industrial Innovation.*
- The Impact of Tax and Financial Regulatory Policies on Industrial Innovation.*
- Antitrust, Uncertainty, and Technological Innovation.*

This report on the electronics industry is one of seven industry-specific studies that were conducted as the second phase of work by this committee. Panels were also formed by the committee to address automobiles; ferrous metals; machine tools; pharmaceuticals; fibers, textiles, and apparel; and civil aviation manufacturing. The objective of these studies was to (1) identify gobal shifts of industrial technological capacity on a sector-by-sector basis, (2) relate those shifts in international competitive industrial advantage to technological and other factors, and (3) assess future prospects for further technological change and industrial development.

As a part of the formal studies, each panel developed (1) a brief historical description of the industry, (2) an assessment of the dynamic changes that have been occurring and are anticipated as occurring in the next decade, and (3) a series of policy options and scenarios to describe alternative futures for the industry.

The methodology of the studies included a series of panel meetings involving discussions between (1) experts named to the panel, (2) invited experts from outside the panel who attended as resource persons, and (3) government agency and congressional representatives presenting current governmental views and summaries of current deliberations and oversight efforts.

The writing of this report was done primarily by Dr. Annette M. LaMond, with assistance from Dr. Robert W. Wilson. Dr. LaMond and Dr. Wilson were responsible for providing the industry

---

*Available from the National Academy of Engineering, Office of the Foreign Secretary, 2101 Constitution Avenue, N.W., Washington, D.C. 20418

and policy research and resource assistance necessary to investigate the issues developed by the panel, as well as producing a series of drafts, based on the panel deliberations, which were reviewed and critiqued by the panel members at each of their three meetings.

Contents

The Competitive Status
of the
U.S. Electronics
Industry

1
The Challenge to U.S. Leadership in Electronics

The security and prosperity of the United States during the remaining years of this century and beyond will depend increasingly on the strength of its electronics industry. Despite its extraordinary growth over the past three decades, electronics remains an emerging industry, promising a continuing stream of technological innovation with enormous international stakes. Based on size alone, electronics already ranks as one of the leading industries in the United States. As shown in Table 1-1, the value of product shipments in the industry's major segments reached nearly $100 billion in 1980, up from $8 billion in 1958. By the mid-1980s the industry is expected to generate over $150 billion in annual sales.[1]

The dollar value of industry shipments, however, is only a partial reflection of the central role of electronics in our society. Education, health care, safety, environment, work activity, recreation, entertainment, life style, and aspirations are all fundamentally enhanced by electronic technology. The microprocessor as an alternative to mental drudgery is transforming society today much as the introduction of the low-horsepower motor did by eliminating physical drudgery.

The national defense also rests increasingly on the edge provided by electronic technology. No longer is numerical strength the sole determinant of a nation's military strength.[2] Military capability in command, control and countermeasure, surveillance, guidance, missile seekers, and communications is determined by the design, manufacture, and application of state of the art electronics technology.

During the past decade, Japan launched an unprecedented challenge to the technological leadership of the United States. In the next decade, the United States will face increasing competition from other countries in eastern Asia and western Europe. Unless the nation rallies to the challenge, the technologically

1

TABLE 1-1 Value of Product Shipments for Major Electronic Industries, 1958-1980

Year	Consumer Electronics (SIC 3651)	Telephone & Telegraph Apparatus (SIC 3661)	Radio & TV Communication Equipment (SIC 3662)	Computing Equipment (SIC 3573)	Test, Control & Medical Electronics[a]	Electronic Components (SIC 367)	Total
1980	$5,385	$10,318	$21,380	$24,350	$11,728	$25,588	$98,749
1979	5,344	8,909	18,592	20,399	10,322	21,649	85,215
1978	5,467	8,003	16,074	15,769	8,706	17,272	71,291
1977	4,731	7,119	14,051	12,673	7,089	14,274	59,937
1976	4,143	4,870	12,078	12,078	5,836	12,153	49,214
1975	3,545	4,500	10,755	8,443	5,165	9,872	42,280
1974	3,892	4,883	9,549	8,668	4,696	10,834	42,522
1973	4,158	4,427	9,253	7,085	4,031	10,499	39,453
1972	3,610	3,974	8,377	6,108	3,249	8,561	33,879
1971	2,893	3,708	7,995	5,116	2,853	5,395	27,960
1970	2,966	3,575	8,454	5,671	2,969	5,794	29,429
1968	3,839	2,550	8,285	4,329	2,557[c]	5,810	27,370
1966	3,798	2,145	6,578	3,040[b]	917[c]	6,160	22,638
1964	2,546	1,696	5,625	2,150[b]	690[c]	4,100	16,807
1962	1,980	1,758	4,564	1,175[b]	617[c]	3,900	13,994
1958	1,330	1,139	1,991	410[b]	324[d]	2,340	7,534

[a]Sum of Control and Processing Equipment, Testing and Measuring Equipment, Nuclear Electronic Equipment, and Medical Electronic Equipment from *Electronic Market Data Book*.
[b]Prior to 1967, figures are from *Electronic Market Data Book*, 1969 edition.
[c]Control and Processing Equipment and Testing and Measuring Equipment only. Prior to 1967, source appears to severely underestimate sales of control and processing equipment.
[d]Testing and Measuring Equipment only.

SOURCES: Data for 1958–1980 from U.S. Department of Commerce, *U.S. Industrial Outlook*, 1982, 1980, 1979, 1976, 1975, 1974, 1969, 1966, 1965, and 1964 editions (Washington, D.C.: U.S. Government Printing Office).

preeminent position of the U.S. electronics industry--historically its chief competitive strength--will gradually be eroded.

Although the United States remains the world technology leader in basic electronics research, the dimensions of the foreign challenge are clearly drawn. Japan now dwarfs its international competitors in consumer electronics and has taken direct aim at strategically important markets in semiconductors, computers, and computer-related equipment. Indeed, the Japanese have been especially successful in closing the technological gap with U.S. firms in advanced computer memory technology--a high-volume and rapidly growing product area in which the United States was dominant only five years ago.

The Japanese challenge in semiconductor technology is particularly ominous. Integrated circuits have taken on enormous strategic importance as they have become larger and larger components of complete electronics systems. Advances in microelectronics are the driving force behind advances in systems technology. At the same time, advances in system architecture design require increasing interaction between the development of software and integrated circuit hardware. While the current strength of the U.S. industry in software is a major asset in integrated circuit and systems competition, this strength will be eroded with further decline in our integrated circuit market share. The U.S. position in semiconductor technology is thus crucial to U.S. competitiveness across the entire spectrum of electronics markets.

Japan's current success in electronics lies in the careful selection and purchase of foreign technology, the speed and extent of its adoption, and the ability to refine purchased technology. To date, the technical accomplishments of the Japanese electronics industry have been the greatest in high-volume products, process engineering, and quality control. For all of its progress in hardware, however, the Japanese industry is widely agreed to lag in the development of software and systems technology. In computers the Japanese position remains a secondary one, despite more than a decade of government-sponsored R&D and industry promotion.

Although knowledgeable observers caution that it would be a mistake to underestimate what the Japanese can do through planning, engineering, and working together on large, long-term projects, some others question the ability of the Japanese to move from their past reliance on imports of foreign technology to a new level of innovation. They cite the societal pressures to conform to group expectations as well as the difficult environment faced by would-be entrepreneurs in Japan.[3] Further, although the quality of primary and secondary education in Japan is very high, funding for higher education has historically stood at low levels.[4]

Graduate training is largely for academic careers, and the system of compensation at major firms offers little incentive to undertake graduate study before entering the work force. Although Japanese firms attempt to make up for these deficiencies by elaborate in-house training programs, the question remains as to whether Japan will produce a sufficient supply of highly trained specialists capable of advanced research.

Answers about Japan's ability to take the technological lead in electronics from the United States must await the outcome of the next years. A strong Japanese effort, however, will undoubtedly be encouraged by Japan's highly favorable domestic environment. The Japanese electronics industry operates from an economic base that affords it significant competitive advantages vis à vis U.S. electronics firms. These advantages derive from the Japanese pattern of high savings and high investment, pride in acquiring technological skills and emphasis on quality, supportive labor relations and work-force quality, and stable, competent governmental bureaucracy. Also favoring the Japanese effort is the clear recognition on the part of the Japanese government and business community that there is an overriding need for innovation and a broad consensus that the national interest requires that major efforts be concentrated in this area.

One of the Japanese electronics industry's greatest advantages in moving to a higher level of innovation derives from the stable, low-interest rate environment that has prevailed in Japan since the end of World War II. In contrast, U.S. interest rates not only have frequently been higher than Japanese rates, but also have moved in a highly erratic fashion. Over the past decade, the interest rate differential between Japan and the United States has widened, ranging between 7 and 10 percent in recent years. Lower interest rates, combined with the high debt-equity ratios permitted by the Japanese financial system, have given Japanese firms a substantial cost-of-capital advantage. This cost-of-capital advantage translates into a significant Japanese production cost advantage--one that has increased as equipment prices have escalated. For example, the Semiconductor Industry Association has estimated that the overall advantage in the production of integrated circuits due strictly to a lower cost of capital, given an assumed cost-of-capital advantage of 50 percent, would be 12 percent.[5] In addition, by lowering project hurdle rates, low capital costs increase the number of projects that can be undertaken. Beyond the cost-of-capital advantage, of course, Japanese firms derive other cost advantages from favorable depreciation schedules and assorted tax credits.

Lower capital costs have thus stimulated a high rate of corporate investment in the Japanese electronics industry. Further, a lower user cost of capital for all Japanese industry has meant relatively more investment generally, which has, in turn, increased the demand for electronics products and spurred induced investment, e.g., in automation and robotics. In a volume-sensitive industry such as electronics, increased demand has not only improved Japan's competitive cost position, but also reduced the risks involved in a technology leap-frog strategy.

The high rate of investment encouraged by low capital costs has been central to the Japanese electronics industry's advances in productivity, product quality, and product development. The effects of this high investment rate can be seen throughout the industry. In consumer electronics, investment in the redesign of television receivers reduced the number of parts in Japanese sets well below the number in U.S. products, with great benefits in terms of both manufacturing cost and product reliability. Similarly, the Japanese were able to sustain their investment in development of video cassette recorders (VCRs) long after U.S. firms had dropped such efforts, and Japanese firms now dominate world VCR production as a result. In the semiconductor industry, the Japanese began to invest heavily in automating the labor-intensive stages of production in the early 1970s, while the U.S. industry looked to offshore production in low wage-rate countries to achieve cost savings. One result was the widely reported difference in Japanese and U.S. product quality in 4K and 16K random-access memories (RAMs). Further, in 1978, when demand for the dynamic 16K RAM overwhelmed the capacity-limited output of U.S. suppliers, Japanese manufacturers, possessed with ample production capacity, moved in to fill the void.

The emergence of the Japanese electronics industry as a major competitor raises a formidable challenge to the technological leadership of the U.S. industry. This challenge comes at a time when U.S. electronics firms face constraints on their ability to invest in new plants and equipment and to sustain the research and development effort needed to stay abreast of rapid technological change. The threat posed by these problems is heightened by protectionist policies that restrict the access of U.S. firms to world markets and, since mid-1980, by the combination of an overvalued dollar and undervalued yen that has in effect created a major barrier to U.S. exports.

The objective of this report is to assess the influences of technology on the international competitive position of the U.S. electronics industry. The number of products, markets, and firms in the U.S. electronics industry is, of course, too large to permit a detailed review of each segment of the industry. Rather, the focus of the study is placed on four broad product groups: integrated circuits, computers, telecommunications equipment, and consumer video equipment. Taken together, these product groups illustrate the dimensions of the problems raised by high capital costs, shortages of technical personnel, and trade barriers to foreign markets.[6]

The strategic role of the integrated circuit industry is high-lighted throughout the report. Integrated circuits have become larger and larger components of complete electronic systems. At the same time, close interaction between integrated circuit manufacturers and systems producers has become increasingly important in designing competitive new equipment. Continued leadership in microelectronic hardware and software technology will thus be crucial to U.S. competitiveness in major equipment sectors such as computers, telecommunications, industrial con-trols and robotics, and consumer electronics in the years ahead. Yet, it is the integrated circuit industry that currently faces the most severe pressures on its human and financial resource base.

The report is organized as follows. The next chapter describes the financial and human resource problems that confront the U.S. electronics industry. Chapter 3 discusses international trade and monetary barriers facing U.S. electronics exports. The next four chapters examine in greater detail the U.S. position and the basis for international competition in integrated circuits, computers, telecommunications equipment, and consumer video equipment. Finally, Chapter 8 provides an overview of the problems and policy options faced by the United States in shaping a competitive policy for the electronics industry.

The policy-related discussions in Chapters 2, 3, and 8 are based on the industry analyses in Chapters 4 through 7. Readers who are primarily interested in policy issues can focus on the introductions to Chapters 4 through 7, which contain summaries of the industry studies. For those readers interested in a particular industry, the industry chapters can be read individually and without extensive reference to other sections of the report.

NOTES

1. See, for example, "U.S. Markets," Electronics, 13 January 1983.

2. See, for example, William J. Perry and Cynthia A. Roberts, "Winning Through Sophistication: How to Meet the Soviet Military Challenge," Technology Review, July 1982, and "Killer electronic weaponry: Tipping the balance in military power," Business Week, 20 September 1982.

3. Although the pressure to conform is often viewed as oppressive to innovation, it can be argued that the supportive group context in Japan allows creative risk-taking because the costs of failure are less extreme than in the more punitive work climate in the United States. It can also be held that the

processes of research and discovery in modern industry are inherently more a group activity than an individual activity.

4. Nevertheless, engineering and science attract many of Japan's top students. In electrical engineering, bachelor's degree graduates increased to 19,572 in 1979, up from 11,335 in 1970, and the enrollments are growing.

5. Semiconductor Industry Association, The International Microelectronic Challenge, May 1981.

6. Major electronics sectors not examined in the report include radio and television communications, and test, control, and medical electronics. Many of the products in these sectors are sold in relatively small and specialized markets. This characteristic lessens the pressure from foreign competition for two reasons. First, it is difficult for foreign (or domestic) firms to achieve substantial economies of scale from mass production. Second, domestic firms possess a competitive advantage based on proximity to customers, particularly in highly specialized product areas, such as test equipment. The degree to which these factors favor domestic firms, however, varies by product and can change over time. For example, foreign competition, particularly by Japanese producers, has become more intense in television broadcast equipment as the industry has moved toward more portable equipment and as program originators at other than the network level adopt equipment with lower cost and transmission quality.

Even in the industry segments analyzed in the following chapters, it has not been possible to address all products and developments of potential competitive significance. For example, one omission in the semiconductor chapter is power circuitry, an area that is central to advances in automation, robotics, and computer control of factory processes.

2
Financial and
Human Resource Constraints

The U.S. electronics industry has had a remarkable record of technological progress, marked by a continual flow of new products and steady advances in manufacturing processes. Over the past three decades, electronic devices have not only replaced mechanical systems and enhanced features in existing products, but also created new products, services, and even industries. At the same time, continued improvement in product performance has been accompanied by dramatic reductions in product cost. The speed per unit cost of a typical IBM computer, for example, increased nearly a million-fold between 1953 and 1979.

This innovative behavior is linked to a unique industry structure that has supported both basic research and subsequent development efforts. The industry's strength in basic research has been sustained by large, progressive firms that possess the resources and incentives to undertake long-term commitments to risky projects with uncertain commercial outcomes. At the same time, smaller, entrepreneurial firms have played a central role, both in introducing new products and in accelerating the pace of innovation by existing firms.

The rapid pace of innovation and diffusion that has given the U.S. electronics industry world technological leadership, however, is threatened by growing financial and human resource pressures. Many observers question whether, in the present investment climate in the United States, internal sources of funds and external capital markets can support the investment in innovation and expansion of production capacity necessary for the U.S. electronics industry to remain a vigorous international competitor. Similarly, shortages of highly trained engineers and computer scientists, particularly on university faculties, may compromise the ability of the industry to design and manufacture state of the art electronic components and systems.

This chapter examines the capital and human resource constraints confronting the U.S. electronics industry. The first section reviews the historical contribution that the unique structure of the U.S. industry has made to the innovative performance of U.S. firms. The following two sections discuss the increasing problems faced by the industry in funding the investment in research, development, and production capacity needed to translate advances in scientific knowledge into marketable products. The final section addresses the issues surrounding the supply of technically trained personnel.

INDUSTRY STRUCTURE AND INNOVATION

The phenomenal growth of the U.S. semiconductor, computer, and telecommunications equipment industries over the past three decades has taken place with little direct government involvement. Rather, the development and diffusion of technological innovations have been encouraged by an industry structure that encompasses both large, vertically integrated firms and a host of smaller, entrepreneurial firms. The U.S. industry, unlike its international counterparts, has historically been characterized by relatively easy conditions of entry. As a result, entry by entrepreneurial firms has accelerated the pace at which advances originating in large industrial and university research laboratories have been translated into commerical innovations.

The innovative contributions made by entrepreneurial firms in the U.S. electronics industry have been founded on the strength of the United States in basic or fundamental research.[1] Fundamental research projects in electronics involve long time horizons, typically running from 5 to 10 years, and uncertain commercial outcomes. To sustain the motivation of researchers over such long time periods, a firm must provide the opportunity to publish interim results and to communicate with researchers at other institutions. The length of time required for a fundamental research project, together with exchange between researchers, increases the probability that research results will diffuse to other firms before commerical returns can be realized.[2] Most fundamental research has been conducted in universities or in firms such as AT&T, General Electric, IBM, and RCA, which are large enough to capture a remunerative portion of the ultimate benefits. Although smaller firms do account for some fundamental research, their share is disproportionately small.

Small entrepreneurial firms in the U.S. electronics industry have nevertheless served to speed the development and diffusion of technological innovation. Entry by such firms has been encouraged by the existence of a well-developed venture capital market. In contrast, in Japan a corporate finance system that relies largely on commercial bank loans, together with the absence of a significant venture capital market, has deterred entry by entrepreneurial firms, producing an industry structure that is dominated by large, vertically integrated firms.[3] Similarly, in western Europe relatively difficult entry conditions have prevented indigenous entrepreneurial firms from assuming the role they have played in the United States.

Entry by new firms in the United States has also been facilitated by a high degree of personnel mobility. Unlike its Japanese and western European counterparts, the U.S. electronics industry is characterized by a pattern of entrepreneurs leaving established firms to form new ventures and of experts being recruited away from a firm by its competitors. Such mobility is particularly rare in Japan where lifetime employment at major firms serves to inhibit "jobhopping."[4]

In the U.S. semiconductor industry, for example, relatively easy entry has promoted the rapid exploitation of new ideas by entrepreneurial firms and has accelerated the adoption of major innovations by existing firms. During the 1960s and early 1970s more than 50 new firms, financed by readily available venture capital, entered the semiconductor industry. The presence of these entrants has been a major force in the industry, encouraging rapid diffusion of technical knowledge and spurring intense competition among industry participants.

New entry has also contributed to the innovative vigor of the U.S. computer and telecommunications industries.[5] From a structure dominated by a small number of mainframe manufacturers in 1960, the computer industry has been transformed by successive waves of new entrepreneurial firms in minicomputers, peripheral equipment, distributed data processing and, most recently, personal computers. The growth in demand for minicomputers a decade ago spawned about 40 new companies; a wave of entry since 1977 has brought an estimated 140 new companies into the industry.[6] Examples of entry in the telecommunications equipment industry are less abundant, but have increased with the reduction in regulatory barriers that began in 1968 with the Carterfone decision. Indeed, in recent years, entry has speeded the development of major innovations such as digital PBXs and communications systems designed for the office of the future.

Advances in electronics technology over the next decade--in areas such as component integration, computer architecture, and opto-electronics--will continue to create opportunities for sustained innovation by both large and smaller specialist firms. The ability of U.S. firms to invest in research, development, and production capacity is a prerequisite for long-run competitiveness. The following sections discuss the capital and human resource problems that must be hurdled by the U.S. industry if it is to remain a vigorous competitor in world electronics markets.

FUNDING FOR RESEARCH AND DEVELOPMENT

The rapid rate of technological change in electronics requires very high levels of R&D spending relative to other manufacturing industries. Many advanced electronics products have lifetimes as short as five years due to technological rather than physical obsolescence. As a result, the ratio of R&D to sales in the computer, semiconductor, and telecommunications equipment industries exceeds 5 percent. Specialized integrated circuit manufacturers devote percentages of sales to R&D that fall in the 8 to 10 percent range.[7] These figures stand in contrast to an average research intensity ratio of 2 percent for all U.S. manufacturing firms.

Research and development costs across the electronics industry have risen rapidly as advances in integrated circuit design and manufacturing techniques have made it possible to put an ever increasing number of transistors on a single chip. Initially, integrated circuits (ICs) were general building blocks that were incorporated into complex logic systems by designers with expertise in network theory, component properties, and practical knowledge of effective design techniques. Software techniques were limited to computers and the minimization of parts in logic system designs.

Greater circuit density has resulted in IC components taking on an increasing number of system functions. Large-Scale Integration (LSI), which began in the early 1970s, fundamentally changed the design process as components assumed large functional blocks of system operation and allowed software design of complex logic systems. LSI made software an integral part of system design and cost. The portion of R&D spent on software development has increased over time and today accounts for over 50 percent of the R&D budget of most computer and telecommunications equipment systems manufacturers. With rising system complexity, few prospects for increased programmer productivity, and continuing shortages of software engineers, it is likely that software costs will continue to escalate.

As device fabrication techniques have improved, Very Large-Scale Integration (VLSI) has become feasible. The complexity of VLSI circuitry, even in more straightforward applications such as RAMs, has resulted in very high design costs. The task of testing and designing software for nonmemory VLSI devices with programmable functions promises to be even more expensive than for the LSI generation products and will require IC producers to undertake an increasing role in software development themselves.

Research and development costs have thus risen rapidly, particularly for complex systems, as the importance of software in product development has grown. High capital costs and shortages of highly trained personnel, however, may be making it more difficult for even relatively large firms to finance long-term commercially oriented R&D. Given these considerations, national governments in many countries are increasingly willing to finance research in electronics.

In part, increased government interest in electronics reflects its growing national security and defense importance. For example, rapid and accurate information transmission is a major determinant of defense performance. Indeed, Pulse Code Modulation (PCM) transmission, microwave and satellite communications, and packet switching were initially developed for military applications.

Governments, however, are also responding to public perceptions of the future role of the electronics industry. It is widely believed that technological developments in these industries will be key factors in future industrial growth. The electronics-based industries are seen not only as increasingly important in terms of output, but also as having vast implications for improving productivity in other industries.

The governments of France, West Germany, the United Kingdom, and Japan have all undertaken major electronics R&D programs in an effort to stimulate innovation and thereby contribute to the international competitiveness of their national industries. These funding programs have been aimed at commerical applications of advanced microelectronics technology, particularly in the areas of computer and telecommunications research. Funding by foreign governments for advanced semiconductor research alone has been estimated at $1.6 billion over the 1978 to 1981 period.[8]

The Japanese government's efforts to encourage innovation in the electronics industry provide a widely noted example of a coordinated national approach to industrial innovation and diffusion. Since the mid-1960s, MITI's Agency of Industrial Science and Technology has sponsored a national R&D program for commercially oriented projects that would not be undertaken by private firms because of the magnitude of the investment required, long-term gestation, and high risk.[9] Under this program, projects are carried out cooperatively by universities, government laboratories, and industry. Projects funded under the national research program include the $400 million, four-year VLSI project, which greatly strengthened Japan's international competitive position.[10] Also of great potential competitive significance is the recently completed Pattern Information Processing System (PIPS) project, which covered both electronic image processing and speech recognition.[11]

The most extensive involvement of the Japanese government in a particular industry, however, involves computers.[12] The statutory basis for these measures is the 1971 law, "Temporary

Measures for the Promotion of Specified Machinery Industries," which authorizes the government to formulate an improvement plan for individual industries. Currently, MITI is launching its latest Super Computer Project, an eight-year effort to coordinate and focus Japanese research in the Josephson junction and in gallium arsenide integrated circuits--two key component areas in which Japan has signalled its intention of topping U.S. technology.[13] Other government efforts to enhance the competitive position of the computer industry are focused on software development, where Japan lags behind the United States. Finally, MITI's recently announced 10-year, $450 million Fifth Generation Computer Project aims to develop a prototype of a new family of machines, designed especially for artificial intelligence applications.[14] Even if the Fifth Generation project falls short of its highly ambitious goals, most observers expect that the focus and consensus provided by the project will give a major boost to the Japanese computer industry.

In addition to the national R&D projects, the Japanese government supports industrial technology through the activities of the national research laboratories, including MITI's Electrotechnical Laboratory and Nippon Telegraph and Telephone's Electrical Communications Laboratory. Special tax concessions also serve as encouragements to industrial R&D. These tax incentives include accelerated depreciation for the construction of research facilities and pilot plants and initial expense of research associations, as well as partial tax exemptions for corporate profits originating in receipts from sale of technology abroad. Such measures, together with MITI's ability to coordinate and influence, have had a far-ranging influence on the international competitiveness of the Japanese industry.

INVESTMENT IN PRODUCTION CAPACITY

Rapidly changing technology calls for a continuing high rate of capital formation within the electronics industry, placing steady upward pressure on capital-output ratios across the industry. At the same time, the risk element associated with investment in the electronics industry is increasing due to greater uncertainty about market shares and profit margins. This uncertainty threatens the ability of the industry to adjust its capital stock to new opportunities.

The magnitude of the requirements for new capital formation in the electronics-based industries is illustrated in Table 2-1, which shows the increase in the ratio of new investments to value of shipments for several electronics industry groups between 1977 and 1980. Although capital requirements have increased across all segments of the electronics industry, the increase has been

TABLE 2-1   New Capital Expenditures as a Percentage of Shipments for Major Electronics Industries, 1977 and 1980

Industry	Industry Shipments (millions of dollars)		New Capital Expenditures (millions of dollars)		New Capital as a Percentage of Shipments	
	1977	1980	1977	1980	1977	1980
Electronic Computing Equipment (SIC 3573)	12,924	26,594	652	1,738	5.0	6.5
Radio and TV Receiving Equipment (SIC 3651)	5,732	6,799	106	159	1.8	2.3
Telephone and Telegraph Apparatus (SIC 3661)	7,858	12,283	217	517	2.8	4.2
Radio and TV Communications Equipment (SIC 3662)	14,886	23,752	471	965	3.2	4.1
Semiconductors and Related Devices (SIC 3674)	5,327	10,501	409	1,546	7.7	15.2

SOURCE: U.S. Department of Commerce, *Annual Survey of Manufactures: Statistics for Industry Groups and Industries*, 1977 and 1980.

most dramatic in the semiconductor industry. Since the end of the 1975 recession, U.S. semiconductor firms have dramatically expanded their production facilities. Further, they have had to retrofit existing facilities with new production equipment since the rate of process equipment obsolescence is exceptionally high. For example, the Semiconductor Industry Association (SIA) estimated that the average age of capital equipment in 1979 for a sample of merchant semiconductor companies was less than five years.

With the evolution of integrated circuits, the increased complexity of necessary production and test equipment has made new IC production capacity very costly. For example, increased density requires high-resolution equipment for transferring IC designs on to silicon. Traditional photolithography techniques are giving way to more costly electron beam and X-ray techniques. In addition, the increased complexity of some circuit designs is dependent on computer-aided design methods, which require more complex hardware and software than that required in the design of earlier, less-sophisticated devices.[15]

These developments, combined with intense competition to increase product performance-price ratios, have resulted in a dramatic increase in the semiconductor industry's capital-output ratio. In the 1960s, $1.00 of capital equipment generated approximately $10.00 of sales. Today, the ratio of capital to new sales is approximately one-to-one. By 1990, $1.00 of new capital investment is expected to generate only $0.50 of new sales.

The capital intensity of systems manufacture has increased less rapidly than that required for IC production. Systems manufacture has historically been an assembly process in which components are placed together on printed circuit boards, and the boards are then connected together and to other devices, such as power supplies and display tubes, within an enclosure, typically a metal cabinet. Testing occurs at various points in the assembly process. The cost of test equipment has increased as IC components have become more complex and now represents a major capital investment for systems manufacturers. In addition to test equipment, capital equipment for plant automation, e.g., for dip soldering and component insertion, has become common to system manufacture.

As capital intensity increases, the struggle for competitive leadership in high-volume electronic products is increasingly being waged on the production line. If U.S. producers are less able financially to undertake necessary investment in plant and equipment, foreign producers are likely to increase capacity to supply increases in demand, and U.S. producers will lose market share. Indeed, the availability of capital required for investment in plant and equipment and the willingness of producers to commit funds even during a downturn in the business cycle can be critical factors. For example, a cutback in investment for plant and equipment by U.S. integrated circuit manufacturers was the principal cause of the industry's capacity problems in the late 1970s. The capacity-limited production of U.S. manufacturers, in turn, provided Japanese producers with the opportunity to increase worldwide market share.

The large investment in production capacity required to support future U.S. growth in the electronics industry is a particular problem for small producers. The cost of necessary growth is more easily financed by large, diversified producers from internally generated funds or external capital markets. In contrast, small and medium-sized producers are likely to have more limited sources of internal funds and face less attractive terms on external funds. Indeed, at a time when interest rates persist at high levels, raising sufficient capital has been a problem even for large electronics firms.

The availability and cost of capital have caused particular problems for the U.S. semiconductor industry, where lower than average profitability has made it difficult to finance increased requirements for capital internally.[16] In the face of intense

competition, benefits from increased automation and more effi-
cient equipment have not resulted in increased profits, but have
been passed on to the customers in the form of lower prices. As
leading producers have placed new advanced equipment into
operation, other producers have followed in order to maintain
market share. Given the volume-sensitive nature of the industry,
increased automation has thus resulted in dramatic price reductions rather than increased profitability.[17]

Finally, the rising capital intensity of the electronics industry
has increased the importance of international differences in
capital costs as a determinant of the competitiveness of U.S.
firms. The lower cost of capital in Japan has proved to be an
important competitive advantage for Japanese firms, particularly
in the semiconductor industry. In 1980, for example, the cost of
capital facing Japanese semiconductor companies was estimated
to be 9.3 percent compared to 17.5 percent for U.S. firms.[18]

The cost-of-capital differential between Japan and the United
States is largely attributable to two factors. First, the institutional structure of the Japanese economy allows Japanese firms
to operate with much higher degrees of debt leverage than U.S.
firms.[19] If U.S. semiconductor firms had a ratio of debt-to-
equity as high as that of Japanese firms, over two-thirds of the
1980 difference in cost of capital would be eliminated. Second,
the cost of long-term debt in Japan is lower than in the United
States, reflecting tightly controlled capital markets, higher
Japanese saving rates, lower inflation, and less restrictive
monetary policies.[20]

Differential capital availability and costs affect the competitive position of U.S. firms in two major ways. First, capital costs
have a direct effect on manufacturing cost that is proportional to
the capital intensity of the manufacturing process. Moreover, to
the extent that a firm enjoys lower capital costs, it will possess a
greater ability to expand production capacity in anticipation of
future demand. Second, capital costs have an obvious impact on
project choice. If foreign firms have a lower cost of capital than
U.S. firms, they will be able to undertake some investment
projects that must be rejected by U.S. firms because the expected
rate of return is below their cost of capital. As a result, foreign
firms facing lower capital costs will tend to build a larger production base and sales volume than U.S. firms. Larger volume, in
turn, may offer an important advantage in funding future R&D.[21]

TECHNICAL PERSONNEL

Electronic products and services embody an unprecedented amount of human knowledge and technically sophisticated labor. Scientists, engineers, and technicians are required for the R&D process, while production engineers and technicians are needed to implement new product and process technology. Selling new products involves technical skills in both the supplier's sales and service force and the customer's organization. Indeed, a large proportion of the electronics industry's sales force is composed of engineering graduates. Similarly, service personnel may range from technicians to engineers, depending on the complexity of the product. In many cases, technically knowledgeable people within purchasing organizations are a prerequisite for the successful commerical introduction of advanced electronics products.

In the late 1970s, the U.S. electronics industry confronted a serious shortage of technical personnel. In part this shortage traced to cutbacks in military and space funding of R&D and procurement that occurred in the late 1960s. Beginning in 1968, real-dollar federal investment in R&D and in science and engineering education, including support for graduate fellowships, began to decline. These declines marked an end to the rapid expansion of university science and engineering departments. In addition, the 1970-1971 recession cut sharply into the demand for engineers. As a result, many students considering engineering switched to other alternatives, and universities trimmed their capacity for engineering education.

These trends are reflected in Table 2-2, which shows the number of electrical engineering degrees awarded at all levels in the United States over the 1969 to 1980 period. The number of bachelor's degree electrical engineering graduates increased until 1972, when the students who entered in 1968, before the cutbacks in government funding, would have matriculated. The number of master's degree and Ph.D. graduates also turned down after 1971 and 1972, respectively. By 1977, the number of electrical engineering degrees granted was at a low point. However, as rapid growth in the electronics industry generated increased demand for engineers following the 1970-1971 recession, enrollments began to increase once more. The number of bachelor's degrees awarded has increased since 1977, and recently the number of master's graduates has increased slightly, though Ph.D. production has continued to decline.[22]

The supply of experienced personnel trained in computer science is also critically short at all degree levels.[23] Trends for computer professionals have been similar to electrical engineering: namely, rising undergraduate enrollments and decreasing Ph.D. production. However, while the number of

TABLE 2-2   Annual Electrical and Electronic Engineering Graduates in the
United States, 1969-1980

Year	B.S.	M.S.	Ph.D.	Total
1969	11,375	4,049	858	16,282
1970	11,921	4,150	873	16,944
1971	12,145	4,359	899	17,403
1972	12,430	4,352	850	17,632
1973	11,844	4,151	820	16,815
1974	11,347	3,702	700	15,749
1975	10,277	3,587	673	14,537
1976	9,954	3,782	644	14,480
1977	9,837	3,674	574	14,085
1978	10,702	3,475	524	14,701
1979	12,213	3,335	545	16,093
1980	13,594	3,660	523	17,777

SOURCE: *Engineering Manpower Bulletin* (U.S.)

graduates in this area increased by over 10 percent per year
during the 1970s, the initial base was small. Thus, despite recent
increases in computer science enrollment, the National Science
Foundation estimated in 1981 that there were 54,000 openings in
the United States for bachelor's degree graduates, but only 13,000
graduates.[24]

Expansion of the supply of well-trained electrical engineers
and computer professionals faces a critical roadblock as a result
of the decline in the number of Ph.D. graduates in the 1970s.[25]
University engineering and computer science departments are
seriously short of both junior and senior faculty, with little
prospect for early improvement. According to the American
Council on Education, more than 1500 faculty positions in the
nation's accredited schools of engineering--10 percent of the
total--were unfilled in the 1980-1981 academic year, and almost
half of those posts had been vacant for more than a year. In
electrical engineering and computer science departments, the
vacancy rate is estimated at 16 percent, with openings in nearly
50 percent of the faculty positions in solid-state electronics,
computer engineering, and digital systems.

One reason for the decline in the number of Ph.D. graduates
and the shortage of engineering faculty is the impressive salaries
and benefits that bachelor's degree engineers have been able to
command in industry in recent years. In contrast, academic
electrical engineering and computer science department salaries
are some 25 to 100 percent below industry levels. The net result

has been a reduction in the ability of universities to provide education in engineering and computer science, although undergraduate demand in these areas is more intense than ever.

Several factors in addition to noncompetitive salaries contribute to the problem of attracting and retaining qualified faculty members. In recent years the attraction of being able to work with graduate students and conduct research in an atmosphere of academic freedom has been tarnished by difficulties in obtaining research support, problems of inadequate equipment and facilities, the instability of government funding for research, and recently by government pressure to restrict scientific publication. Further, the current shortage of graduate students and faculty members creates unusually heavy teaching loads, which make academic jobs less attractive for those interested in research.

An important additional problem in engineering education is a severe lack of the equipment required for instructional purposes. Much of the laboratory equipment and physical facilities being used for both teaching and research purposes in university engineering and computer science departments was acquired during the 1960s. The obsolescence of this equipment implies that part of the education that undergraduate engineers and computer scientists receive is itself obsolete vis à vis current industrial practice. For example, the apparatus needed to teach computer-aided design/computer-aided manufacture (CAD/CAM) methods-- the source of important productivity gains for some large corporations over the past decade--is generally unavailable in engineering schools. While this situation may not pose an insurmountable problem for a larger employer who can afford on-the-job training for new personnel, it may have adverse effects on smaller companies that depend upon new graduates for information about the latest developments in engineering practice. Recently, the situation has improved somewhat with graduate research centers initiated through a combination of government and industrial support. However, much of the ground lost during the 1970s must now be regained at sharply higher cost.

Accelerating change in electronics technology has also created problems in continuing education and retraining of engineering professionals. Engineering technology is advancing so rapidly that the half-life of a professional's knowledge base is now estimated at two to five years. An engineer whose only continuing education consists of occasional brush-up courses does not have a chance of staying technically current. At present, however, continuing education for engineers is spread among a variety of sources, including industrial firms, consultants, professional societies, and colleges and universities, with little or no coordination. There has been virtually no federal support for continuing education, in part because the costs of industrial programs have been regarded as business expenses.

TABLE 2-3   Electrical Engineering Graduates per Million Population

	1965	1970	1975	1977
France	20	34	28	33
Japan	82	133	162	185
United Kingdom	32	46	45	46
United States	–	85	67	66
West Germany	16	11	48	109

SOURCE: Semiconductor Industry Association, *The International Microelectronic Challenge*, May 1981.

There is currently considerable concern about the relative trends in the education of engineers and computer professionals in the United States as compared with other highly industrialized countries. Japan and West Germany are producing much larger proportions of engineers than the United States, as indicated in Table 2-3.[26] At the same time, these countries are educating a substantial majority of their secondary school populations to a point of considerable scientific and mathematical literacy. In contrast, the United States has placed decreasing priority on science and mathematics for secondary students not intending to major in science or engineering.

Japanese graduates in electrical engineering at all degree levels have increased dramatically in the postwar period, as shown in Table 2-4.[27] Indeed, the number of Japanese graduates in electrical engineering surpasses the U.S. level both on a per capita basis and in absolute terms, even though Japan has only one-half the population base of the United States. The rapid increase in Japanese engineering enrollments, however, has been criticized by scholars of higher education for having been accomplished by increasing student-faculty ratios and economizing on university resources per student.[28] Moreover, Japanese engineering education is highly abstract with little or no practical laboratory experience. Nevertheless, two features of the Japanese system may mitigate the effect of educating large numbers at low cost in the universities: the rigor of high school instruction in mathematics and the post-university training that large companies provide for their engineers.[29]

Large Japanese firms invest heavily in training newly graduated engineers given the incentives provided by the lifetime employment system. Indeed, the Japanese investment in training is known for its elaborate character, including in-house training programs, company technical institutes, university graduate

TABLE 2-4   Annual Electrical Engineering Graduates in Japan, 1969-1979

Year	B.S.	M.S.	Ph.D.	Total
1969	11,335	705	108	11,848
1970	13,085	688	116	13,889
1971	14,361	844	109	15,165
1972	16,020	913	119	17,052
1973	16,025	1,026	114	17,345
1974	16,140	1,173	106	17,419
1975	16,662	1,258	120	18,040
1976	16,943	1,201	114	18,258
1977	17,868	1,447	142	19,257
1978	18,308	1,686	132	20,126
1979	19,572	1,697	166	21,435

SOURCE: Ministry of Education (Japan).

education, as well as rotation through a variety of jobs.[30] In the context of lifetime employment, such training yields benefits by reducing both recruitment costs and resistance to technological change and job reassignment.[31]

The U.S. electronics industry also faces shortages in the number of skilled technicians who provide support for engineers and computer science professionals. Other industrialized countries such as Japan and West Germany place heavy emphasis on training technicians in special vocational schools. These countries not only offer technicians good jobs but also considerable social status as well. In contrast, technician training in the United States has largely been a haphazard enterprise, accomplished through a combination of on-the-job training, a few technical institutes, vocational training in secondary schools, and high-grade, very high-cost educational programs in the armed forces. Although two-year community colleges have begun recently to play an important role in technician training, these colleges share a number of problems with the higher education system with respect to faculty and equipment.[32]

Finally, the basic scientific and mathematical education of all U.S. citizens is at issue. There is a growing discrepancy between the science and mathematics education acquired by high school graduates who plan to follow scientific and engineering careers, and those who do not. Although scientific and technological literacy is increasingly important in our society, more students than ever before are dropping out of science and mathematics courses after the tenth grade, and this trend shows no signs of abating. This trend has troubling implications not only for the size of the future pool of technical personnel but also for the ability of the United States to generate and incorporate technological change in its production and utilization of goods and services.

NOTES

1. The term "basic research" as defined by the National Science Foundation covers "projects which represent original investigation for the advancement of scientific knowledge and which do not have specific commercial objectives, although they may be in fields of present or potential interest to the reporting company." Many firms thus prefer the term "fundamental research," i.e., research on difficult problems, often with long-run horizons and uncertain outcomes, but in some cases specific commercial objectives, to describe their research activities.

2. Communication and language barriers generally make diffusion of new knowledge between countries slower than diffusion within a country.

3. MITI, however, has recently proposed new measures to make it easier for new ventures to raise capital. Further, foreign companies are increasingly interested in promoting new Japanese enterprises. See, for example, "Japan: Smoothing the way for venture capital," Business Week, 11 October 1982.

4. Nevertheless, the number of individuals leaving large Japanese electronics firms to set up their own companies has increased in recent years. See, for example, "The exodus that shook the establishment," Business Week, 14 December 1981.

5. A major factor contributing to entry and innovation in these industries in recent years is the low cost and flexibility of off-the-shelf components, such as microprocessors, memories, and peripheral devices. By buying standard parts from manufacturers that sell them in large volumes, new ventures can take advantage of both economies of scale and new technology without incurring the full burden of R&D costs. Cheaper computing power has also encouraged startups by paving the way for the development of more standard software. The combination of a unique architecture and software can be used to create systems that solve generic problems, address special uses such as portability, or meet the needs of particular users.

6. "Computers: The incredible explosion of startups," Business Week, 2 August 1982.

7. "Annual survey of corporate R&D expenditures," Business Week, various years. The lower ratio of R&D to sales for systems companies compared to integrated circuit firms reflects in part the higher level of sales revenue required for sales and service activities in end-user markets.

8. For example, the French government is launching a four-year project to give selected microelectronics companies and laboratories $500 million in funding for R&D. At the same time, the government is increasing its spending for research in data processing and computers to $350 million in 1982, up from $270 million in 1981. In telecommunications, the government-financed research has led to major achievements in widening applications for electronic digital switching. See, "Data processing: Pitfalls in France's vast R&D plan," Business Week, 23 November 1981; "France spends billions on goal of becoming a leader in technology," The Wall Street Journal, 14 September 1982; U.S. International Trade Commission, Competitive Factors Influencing World Trade in Integrated Circuits, Washington, D.C., November 1979; and Semiconductor Industry Association, The International Microelectronic Challenge, May 1981.

9. Selection of national research projects begins with suggestions from companies, universities, and government laboratories; further modifications are made by MITI. Final approval is given by an agency advisory council comprising representatives from the universities, industry, and public. Projects are managed by an agency development officer, with various tasks assigned to companies, government laboratories, and universities. Each is reimbursed its full costs, with companies often further benefiting from a three-year, nonexclusive royalty discounted or reduced license of patents resulting from the projects.

Although all research results are published, the national R&D projects benefit the participating companies far more than others. While outside companies can license the patents, which are held by the government, participating companies pay less for them. Moreover, an outside company that wants to duplicate any patented development cannot obtain the essential details from the company that did the original work.

A second separate agency program provides subsidies for other industrial research that is considered important to the economy, but unlikely to be carried out without government assistance. Proposals originate with companies, who pay half the development cost. Patents, designs, and research results belong exclusively to the developing company. However, this program is relatively small.

See Merton J. Peck, "Technology," in Asia's New Giant, Hugh Patrick and Henry Rosovsky, eds., The Brookings Institution, Washington, D.C., 1976.

10. Of the $400 million funding for the VLSI project, two-thirds consisted of government loans to be repaid from profits realized through the application of VLSI technology developed through the program and one-third of government subsidies. In

addition, MITI provided a technical laboratory where researchers from the five leading Japanese semiconductor firms worked together to develop the new materials and production techniques necessary for VLSI circuitry. Over 700 patents were generated by the project. Nevertheless, some observers credit the specific MITI VLSI project work less than the national consensus it provided in response to which Japanese firms did the job internally. See, for example, U.S. International Trade Commission, Competitive Factors Influencing World Trade in Integrated Circuits, Washington, D.C., November 1979, and Semiconductor Industry Association, The International Microelectronic Challenge, May 1981.

11. Under the PIPS project, MITI's Electrotechnical Laboratory developed the basic technology of nonkeyboard computer input devices, and applications were carried out by Fujitsu, Hitachi, Mitsubishi, NEC, and Toshiba. See, for example, "Electronics research: A quest for global leadership," Business Week, 14 December 1981.

12. In the 1960s, the U.S. computer industry exerted a major demand-pull influence on U.S. semiconductor technology. When the Japanese computer industry failed to perform this role, the Japanese government moved to speed progress in the computer industry and simultaneously push faster development of related semiconductor technology.

13. The Super Computer Project is an "entrusted project," meaning that the government will bear the total expense. See, for example, "Electronics research: The quest for global leadership," Business Week, 14 December 1981; "Japan's super computer push," The New York Times, 28 October 1982; and "With stakes high, race is on for the fastest computer of all," The New York Times, 1 February 1983.

14. "A fifth generation: Computers that think," Business Week, December 1981; and "West wary of Japan's computer plan," Electronics, 15 December 1981.

15. Increased product sophistication, coupled with inflation, has also driven IC production equipment prices up sharply in recent years. From 1975 through 1980, the price of a wafer fabrication module increased at an annual compound rate of nearly 40 percent and is expected to continue at that rate through 1985. The price of a mask aligner, nearly $500,000 in 1981, is expected to reach $1 million in the next few years. By 1980, a typical semiconductor production line, which required an investment of only $1 million in 1965, cost as much as $50 million. U.S. Department of Commerce, U.S. Industrial Outlook 1981, Chapter 26; and "Rolling with recession in semiconductors," Business Week, 21 July 1980.

16. Capital problems, however, are not limited to the semi-conductor industry. See, for example, "Computers: A capital crunch that could change the industry," Business Week, 23 March 1981; and "Moving away from mainframes: The large computer makers' strategy for survival," Business Week, 15 February 1982.

17. Indeed, the poor financial performance of the U.S. semi-conductor industry and depressed conditions in the U.S. stock market have led to a growing foreign interest in the industry. Acquisitions of U.S. semiconductor firms by foreign systems manufacturers since 1975 include: the U.S. Philips Trust's acquisition of Signetics, Siemens' acquisition of 80 percent of the equity of Litronix and 20 percent of the equity of Advanced Micro Devices, Robert Bosch's purchase of 25 percent of the equity of American Microsystems, and Schlumberger's acquisition of Fairchild Camera and Instrument.

18. Chase Financial Policy, U.S. and Japanese Semiconductor Industries: A Financial Comparison, as cited in Semiconductor Industry Association, The International Microelectronic Challenge, May 1981.

19. U.S. semiconductor firms have debt-to-equity ratios of less than 25 percent on average, while four of the Japanese semiconductor firms analyzed in the Chase study maintained debt-to-equity ratios of 150 to 230 percent. In contrast to U.S. firms, Japanese semiconductor firms are able to employ high leverage ratios because of their affiliation with large industrial groups, Japanese lending practices, and a supportive government policy. Chase Financial Policy, op. cit.

High debt-to-equity ratios have allowed Japanese firms to lower their producton costs and, therefore, increase their price competitiveness. Following years of booming sales, however, some of Japan's major electronics manufacturers are relying less on bank borrowing to finance their growth and more on long-term bonds and even common stock. As Japanese firms strengthen their financial positions, they will be able to take greater risks, e.g., in research funding, than they have been able to in the past. See, for example, "How Japan will finance its technology strategy," Business Week, 14 December 1982; and "Japan's latest corporate advantage," The Wall Steet Journal, 16 September 1982.

20. One of the important institutional factors responsible for the high savings rate in Japan is the special treatment of taxation on capital income. Small savers are encouraged through full exemption of taxes on interest income up to the first 9 billion yen (approximately $40,000). Favorable tax treatment is also accorded the capital income of large savers, which can be separated from labor income and is taxed at a maximum rate of 35 percent on interest and dividend income and 16 percent on some bonds. Further, taxation on capital gains is virtually zero in Japan.

21. Larger sales volume provides the potential for a higher level of cash flow with which to finance R&D projects. Internally generated funds and stockholders' equity are used for nearly all R&D projects. One reason for this is that the output of R&D is intangible and, in contrast to capital equipment, does not provide physical collateral for a loan or bond issue. Another reason is that R&D projects are typically more risky than capital investment projects and, therefore, better suited to financing by equity and internal funds than debt.

22. It should be noted that one-third of all engineering Ph.D. candidates are foreign nationals, two-thirds of whom are in the United States on student visas. Although many of those in the latter category may remain in the United States, the total number of new Ph.D.s who enter the labor force each year will be less than the number who receive their degrees. Science and Engineering Education for the 1980s and Beyond, prepared by the National Science Foundation and the U.S Department of Education, October 1980.

23. Entry-level computer programmers, however, are now in oversupply as a result of the dramatic increase in the number of people retraining for computer programming jobs. See, for example, "Jobs for programmers begin to disappear," Business Week, 16 August 1982.

24. Faced with a continuing shortage of computer programmers, computer firms are working on ways to streamline and simplify the programming process. See, for example, "Computer companies develop devices to ease programming," The Wall Street Journal, 25 June 1982.

25. See, for example, National Science Foundation and U.S. Department of Education, Science and Engineering Education for the 1980s and Beyond, October 1980; Pat Hill Hubbard, Plan for Action to Reduce Engineering Shortage . . . . With Supporting Data, American Electronics Association, Palo Alto, California, October 1981; and James Botkin, Dan Dimancescu, and Ray Stata, "High Technology, Higher Education, and High Anxiety," Technology Review, October 1982.

26. In considering cross-national comparisons in engineering, one must be cautious because educational systems are not parallel and may be quite dissimilar. For example, the group labeled "engineers" in one country may include an unknown number of those termed "technicians" in the United States.

27. In Japan, 20 percent of all bachelor's degrees and about 40 percent of all master's degrees are granted to engineers, and these figures have been stable for the past decade. This compares with a figure of about 5 percent at each degree level in the United States.

28. Merton J. Peck, "Technology," in Hugh Patrick and Henry Rosovsky,(eds.), Asia's New Giant, The Brookings Institution, Washington, D.C., 1976.

29. For example, the size of the electrical engineering faculty at Tokyo University has expanded only slightly over the past two decades, and its physical plant has changed little since the 1950s. See, "Japan's strategy for the 80's: Why industry must step in to train engineers," Business Week, 14 December 1981.

30. For example, at Japan's largest consumer electronics company, a design engineer's training includes a tour of duty in the company's retail outlets, selling and servicing products, followed by an assignment in the factory, working on the assembly lines. When he is finished, he has a "personal appreciation" for what the public wants and, presumably, for how best to design a product for producibility and reliability. "American manufacturers strive for quality--Japanese style," Business Week, 12 March 1979.

31. Although it is difficult to assess the value of Japanese training practices, it is likely that the Japanese engineer is more familiar than the U.S. engineer with the capabilities and problems of his own company. On the other hand, the Japanese engineer may be less familiar with practices in other companies and university research in his specialty. Some observers thus believe that the Japanese career pattern is better suited for applying and improving technology, which often involves close working relationships with production engineers, than for major technological innovation.

32. In addition, two-year community colleges also face some unique problems: low faculty retention rates; increased competition from comprehensive colleges and universities searching for students; decreased funding from the local sources on which the colleges heavily rely (financing of two-year colleges comes heavily from local tax sources); and poor preparation in mathematics on the part of students, which absorbs considerable resources for remedial teaching.

3
Barriers to International Trade

In recent years, U.S. electronics firms have increasingly looked to foreign markets, and a greater number of foreign firms have begun to compete in the United States. With rising investment requirements for R&D and production capacity, success in world markets has become an important determinant of a firm's long-run commercial viability. Maximum access to foreign markets is necessary to keep production costs as low as possible and to recover large R&D costs associated with product development. Further, markets that are sheltered give producers located in them the opportunity to engage in price discrimination, e.g., by setting high domestic prices to fund R&D and low prices elsewhere to gain export market share. Increasing trade barriers to exports in Europe, Latin America, and Asia have thus become a serious concern for U.S. electronics manufacturers.

This chapter provides an overview of the international trade problems faced by the U.S. electronics industry. A more detailed discussion of particular problems in major industry segments is contained in Chapters 4 through 7. The first section of the chapter contrasts the strong historical contribution of electronics to the U.S. merchandise trade balance with the problems posed by the current overvaluation of the U.S. dollar in world currency markets. The extraordinary global strength of the dollar since 1980 has in effect created a barrier to U.S. exports--one that goes far in explaining the recent escalation of protectionist sentiment in the United States. The second and third sections examine some of the tariff and non-tariff barriers that confront U.S. firms seeking to gain access to foreign markets. The chapter closes with a note on the need for major reform of the international monetary system. Unless the dollar's value comes down, U.S. electronics exports will be handicapped regardless of the U.S. success in eliminating barriers to trade.

U.S. ELECTRONICS INDUSTRY TRADE BALANCE

The U.S. electronics industry has historically made a substantial positive contribution to the U.S. merchandise trade balance. By 1980, for example, the combined trade balance in the four major electronics industry product groups shown in Table 3-1--electronic components, consumer electronics, electronic computing equipment, and telephone and telegraph equipment--was a $3.3 billion surplus. In contrast, the overall U.S. merchandise trade balance was a $27.7 billion deficit.

The U.S. trade position, however, varies across segments of the electronics industry. One sector--radios and television receiving sets--has shown large deficits, particularly with the Japanese. Electronic components have shown surpluses, despite extensive offshore assembly operations of U.S. firms and the recent increase of Japanese market share in computer memories. Electronic computing equipment has contributed large trade surpluses. As with the semiconductor industry, the computer industry operates an extensive international network of assembly and manufacturing plants. As a result, exports do not fully portray the strength of the U.S. industry's competitive position. Trade in telecommunications equipment has also produced consistent surpluses, though these have been moderate, particularly in relation to U.S. production.

In 1981, the size of the U.S. electronics trade surplus declined due largely to the combination of a strong U.S. dollar and a weak Japanese yen. The unanticipated persistence of high interest rates in the United States during the recent recession, coupled with political and financial uncertainties elsewhere, kept the dollar strong, leading to a decline in U.S. price competitiveness. At the same time, restrictive Japanese capital market procedures served to keep Japanese interest rates low and the yen weak.[1] As a result, Japanese exports gained a substantial price advantage and increased share in foreign markets. As shown in Table 3-2, for example, the Japanese trade balance with the United States improved in each of the industry segments noted above.

If the Japanese system of formal capital market controls and informal guidance were completely relaxed, many observers believe that Japanese interest rates would rise, and the yen would appreciate. However, all agree that result would take time. There is no money market in Japan and only a limited range of government securities. Further, many banks, the major source of funds for the bond market, insist that the bonds they buy be secured by collateral. Thus, in the transition period, greater internationalization of the yen is likely to weaken it further, as Japanese investors continue to pursue the opportunity for higher returns abroad and foreign borrowers of low-interest yen convert their loans into dollars.

TABLE 3-1 U.S. Trade With the World in Electronic-Based Products, 1975–1981 (millions of current dollars)

SIC No.			1975	1976	1977	1978	1979	1980	1981
3661	Telephone and Telegraph Equipment	Exports	198	227	257	388	448	557	653
		Imports	93	100	129	233	319	421	496
3671-9	Electronic Components	Exports	1,987	2,532	2,682	3,006	3,946	4,988	5,166
		Imports	1,160	1,645	2,018	2,676	1,562	4,363	4,935
3651	Radio and Television Receiving Sets	Exports	392	498	467	756	801	1,107	1,045
		Imports	1,810	2,981	3,599	5,039	4,864	4,919	6,503
3573	Electronic Computing Equipment	Exports	2,229	2,588	3,264	4,128	5,389	7,468	8,493
		Imports	129[a]	235[a]	253[a]	755	969	1,159	1,647

[a]Parts for SIC 3573 and 3574 imports were not broken out separately before 1978, but were included in a SIC 3579 category.

SOURCE: Bureau of Industrial Economics, U.S. Department of Commerce.

TABLE 3-2 U.S. Trade with Japan in Electronic-Based Products 1975-1981 (millions of current dollars)

SIC No.			1975	1977	1978	1979	1980	1981
3661	Telephone & Telegraph Equipment	Exports	5.6	4.4	6.3	8.1	8.6	8.2
		Imports	25.6	44.9	92.9	105.6	163.3	247.1
3671-9	Electronic Components	Exports	109.1	133.0	172.3	253.4	238.2	261.5
		Imports	161.2	371.5	502.0	698.6	830.0	962.2
3651	Consumer Electronics	Exports	22.1	25.2	37.2	55.4	61.6	49.9
		Imports	1,251.5	2,047.7	2,757.7	2,350.7	2,337.1	3,646.1
3573	Electronic Computing Equipment	Exports	189.3	279.5	331.1	441.1	607.3	717.6
		Imports	32.3	57.9	187.2	195.7	189.3	386.5

SOURCE: Prepared by Bureau of Economic Analysis, U.S. Department of Commerce.

Sharp improvement in the Japanese trade position and deterioration in that of the United States and western Europe have increased trade tensions worldwide. National governments now face intense pressure to support their domestic producers by restricting access to their home markets. For example, France recently imposed a series of stiff administrative measures to stem foreign imports ranging from video tape recorders to heavy machinery. The following two sections describe some of the major tariff and non-tariff barriers faced by U.S. electronics producers.

TARIFF BARRIERS

Tariffs remain a barrier to the free flow of electronics products in world markets, particularly in the European Economic Community (EEC), where the latest multilateral trade negotiations failed to achieve a reduction in duties. Tariff rates facing U.S. electronics exports in most major foreign countries exceed the corresponding U.S. duty. In 1981, for example, U.S. tariff rates on integrated circuits stood at 5.8 percent, while the corresponding rates facing U.S. exports ranged from a 10 percent ad valorem rate imposed by Japan to the 17 percent common external rate of the EEC to rates as high as 30 percent in some Asian countries.[2]

High tariffs are most burdensome for products where the local industry has reached an essentially equal level of technology. In such cases, a substantial tariff duty either reduces the price competitiveness of the product or forces the exporter to absorb the added cost. Where no equivalent local product exists, high-technology products are not usually excluded by high tariffs. Indeed, the EEC selectively suspends duties on products for which production is inadequate or nonexistent in its member states. Once production is adequate, however, the suspension is lifted and U.S. exports are then placed at a competitive disadvantage. The technological edge possessed by the U.S. electronics industry is thus an essential factor in successful export sales.

A further tariff-related barrier impeding U.S. exports to Europe is created by the "rules of origin" governing nonagricultural trade between the EEC and the European Free Trade Association (EFTA). These rules limit the amount of imported material an item may contain and still receive duty-free treatment when it is exported within the EEC/EFTA area. The rule applied to electronic equipment limits the value of imported components in a finished product to less than 5 percent. This limitation discourages European electronics systems producers from using imported semiconductor devices since it can restrict their potential market or increase their selling cost in other European countries.

In Japan, U.S. exports face tariff barriers related to Japanese customs valuation practices. If Japanese customs officials feel that the invoice value of imported products does not reflect normal value in trade, they raise this value for tariff duty purposes. This practice, called "customs uplift," is alleged by U.S. industry observers to be arbitrary and inadequately explained, as well as time-consuming and expensive.

NON-TARIFF BARRIERS

Non-tariff barriers in Europe and Japan pose an increasingly serious problem for U.S. electronics firms. As described below, these barriers arise from many sources, including restrictions in government procurement, "buy-national" attitudes in private industry, obstacles to the establishment of foreign subsidiaries, standards, closed-end market and distributor practices, import licensing, financial controls, customs procedures, documentation procedures, and various non-tariff charges.[3]

Restrictions in Government Procurement

Restrictions in government procurement affect all segments of the electronics industry, but have their most serious impact on the telecommunications equipment industry. However, the refusal of foreign state-owned telecommunications agencies to use foreign-made equipment or equipment containing significant quantities of imported components also has a negative impact on the integrated circuit industry because the use of integrated circuits in telecommunications equipment is large and growing rapidly.

Barriers to sales of U.S.-produced equipment posed by foreign post telephone and telegraph (PTT) authorities are twofold. Approval procedures for equipment purchases by these agencies, mainly in developed countries, are designed to limit the use of imported equipment. In addition, standards and specifications are geared to locally produced equipment and are often arbitrary, undefined, or unavailable. These difficulties reinforce prevailing policies in Japan and western Europe that require local sourcing of equipment and components, unless they are not available locally.

None of the European PTT authorities is subject to the government procurement code negotiated during the Tokyo Round under the General Agreement on Trade and Tariffs (GATT). In

France, for example, telecommunications equipment procurement is reported to be restricted to a preferential list of suppliers that gives highest priority to locally owned companies, followed by foreign subsidiaries located in France, followed in decreasing preference by the extent of French ownership. Until 1981, the Japanese PTT authority, Nippon Telegraph and Telephone (NTT), also refused to purchase foreign-made telecommunications equipment or equipment using foreign components. However, as a result of bilateral negotiations with the United States, NTT's procurement has now been made more accessible to foreign suppliers. To date, however, NTT procurement from foreign sources remains an infinitesimal portion of the utility's annual budget.[4]

Private Industry Purchases

Reluctance or refusal by private companies in foreign countries to purchase imported electronics products is often cited as a serious non-tariff barrier facing U.S. producers. In some cases, such reluctance is attributable to the difficulties created by other trade barriers, such as foreign exchange controls, import licensing, and standards barriers.

Such restrictions are found in many countries, but they are generally regarded as most serious in Japan. Indeed, a "buy national" psychology is said to pervade much of Japanese industry. For example, in recent submissions to the U.S. International Trade Commission, four major U.S. semiconductor manufacturers reported that Japanese firms that use and distribute integrated circuits will generally purchase imported components only when they are unavailable from a Japanese supplier.

Obstacles to Establishment of Foreign Subsidiaries

Manufacturing and sales subsidiaries in foreign countries are generally viewed as essential to assure access to the market. A formal sales presence in a country is useful in providing technical service as well as in assuring customers with respect to service, product reliability, and delivery. In addition, foreign manufacturing subsidiaries may be necessary to overcome such barriers as tariffs, buy-national policies, and attitudes on the part of public and private enterprises, or any competitive edge based on low local factor costs. Moreover, as discussed above, access to important foreign PTT markets also requires local production facilities in many cases.

Foreign government policies that prohibit or impede direct investment in subsidiary firms or joint ventures may thus pose a significant non-tariff barrier. Difficulties in establishing sales and manufacturing subsidiaries in Japan have been particularly troublesome for U.S. firms. Until 1975, the Japanese, with rare exceptions, effectively prohibited both the construction of foreign-owned plants in Japan and foreign investments in existing Japanese companies.[5] A measure of the effectiveness of Japan's protectionist policies with respect to foreign investment is that, while U.S. semiconductor firms had established 46 subsidiaries in Europe by 1974, including 18 manufacturing operations, only one U.S. semiconductor firm possessed a manufacturing operation in Japan. Even now, establishment of foreign-owned operations in Japan remains difficult and, therefore, will continue to limit U.S. exports.

Standards

Technical product standards can also pose trade barriers in cases where they are difficult to identify or comply with. Standards have frequently been cited by U.S. producers as a significant barrier to exports to Japan.[6] For example, foreign producers are not allowed to participate in the formulation of Japanese industry standards and are generally not notified in a timely fashion or convenient manner of the nature of changes in mandatory standards. In addition, provision for overseas testing of components destined for Japan is lacking, and application of standards to imports is uneven. NTT standards for electronic equipment and components are often based on design, rather than performance, placing outside suppliers, not well acquainted with NTT design specifications, at a disadvantage. Moreover, equipment development for NTT is conducted by local suppliers, making some of the designs proprietary information and closing the contracts to all other suppliers.

Import Licensing

Another potential non-tariff barrier involves import licensing. If licenses are not automatically granted, the costs, delays, and uncertainties associated with compliance can detract from the competitiveness of imported goods relative to domestic products. Brazil is often cited as an example of a country where complex and time-consuming import licensing procedures constitute a barrier to trade. In addition, licensing may be used to administer quantitative restrictions on imports, a practice that is followed mainly in developing countries.

Financial Controls

Foreign exchange controls and other regulations on the method or timing of payment for imports make trade more difficult by increasing the administrative and financial cost of selling overseas. Required prior deposits of up to several times the value of the imported shipment are, in effect, a tax on the transaction, tying up working capital without payment of interest and exposing the deposit to adverse exhange-rate changes during the months that the deposit is held. Such practices are most common in developing countries, mainly for monetary reasons.

Customs Procedures and Documentation

The documentation required of importers in order to take possession of merchandise at point of entry varies from country to country. The industrialized western countries generally have few requirements, while Japan and most developing countries require more complex documentation. Although documentation is not generally processed in a discriminatory manner, complex requirements and slowness of procedures in some countries can cause delay and expense that detract from the competitiveness of imported goods.

Complex and time-consuming customs documentation requirements and procedures in Japan are viewed as a significant trade barrier by U.S. exporters. The Japanese Customs Tariff Law not only involves extensive paperwork and delays, but procedures vary by location. In addition, Japanese customs officials may impose customs clearance charges in excess of import duties, commodity taxes, and documentation fees.

OPENING MARKETS TO INTERNATIONAL COMPETITION

Escalating trade tensions between the United States, Japan, and the EEC have not only frustrated efforts to liberalize existing trade restrictions, but they have also led to a proliferation of new trade barriers. Tariffs--the main bargaining target in past general trade liberalization efforts such as the Tokyo Round--are no longer the major obstacles to trade. Non-tariff barriers have now become the main trade deterrents, and they are far harder to eliminate because their impact on trading partners varies greatly, leaving less room for multilateral trade-offs. For example, the United States is substantially open to foreign competition. In contrast, U.S. companies face formidable trade barriers in Japan. Japanese officials have long resisted U.S. demands that the Japanese market be opened to U.S. exports, arguing that such demands ignore Japan's fundamental vulnerability in an increasingly troubled world economy. Similarly, the EEC is far less anxious than the United States to lower trade barriers. Europeans fear that their domestic electronics industries will be wiped out by more advanced U.S. and Japanese rivals if the EEC agrees to curb subsidies and open its markets to more foreign competition.

Efforts to resolve the current trade conflict without fundamental reform of the international monetary system will meet with little success. Progress toward freer world trade must address the underlying causes of the repeated, and severe, exchange-rate misalignments that have periodically emerged between the dollar and the yen and other currencies. There is no inherent reason why these problems cannot be resolved effectively. It is essential, however, to move quickly to begin the process of adjustment if the mounting pressure for protection in the United States and elsewhere is to be headed off.

NOTES

1. Since late 1980, the spread between U.S. and Japanese interest rates has typically ranged between 7 and 10 percent. In May 1982, for example, the prime rate in the United States was 16.50 percent; its equivalent in Japan was 6.65 percent. Such interest rate differentials are generally seen as evidence that Japan's capital markets are closed. If Japanese money markets were unrestricted, foreign borrowers would, in theory, raise money in Japan to take advantage of the lower cost of funds and, thereby, force Japanese interest rates higher.

Despite slow and steady liberalization over the past decade, Japan's capital markets remain the most tightly controlled of any major economy. In addition to fixing the price of all government bonds--at artificially low rates according to most observers, the Ministry of Finance is also empowered to set limits on the amount of foreign currency domestic and foreign banks can convert into yen and to control the number and size of foreign bond offerings in Japan--so-called samurai bonds. Further, through a regular system of informal consultations, the Ministry of Finance controls the volume and price of overseas loans by Japanese banks and the purchase of foreign securities by domestic insurance and security firms.

Until recently, decisions on which foreign firms could borrow in Japan were determined by strict rules set down by Tokyo's Capital Market Committee, which represents Japan's banks and security houses and is strongly influenced by the Ministry of Finance. In addition, foreign firms seeking money in the Tokyo market had either to prove they were financing specific Japanese exports or to sell yen-denominated bonds to Japanese institutional investors. The bond market was in effect closed to all but the most creditworthy companies offering large issues, $50 million or more, and confined to just one foreign issue each quarter.

See, for example, "Is Japan holding the yen down?" Business Week, 8 March 1982; "Are the Japanese rigging the yen?" Fortune, 31 May 1982; "Japan's capital market has U.S. critics," The New York Times, 1 June 1982; "Borrowing yen will be a little bit easier," Business Week, 31 May 1982; and "Borrowers are eager to get yen loans but must grapple with Japan's delays," The Wall Street Journal, 7 July 1982.

2. In the latest round of Multilateral Trade Negotiations, Japan and the United States established a formula designed to harmonize reciprocal duties on semiconductors by 1987. In that year, a final rate of 4.2 percent will be achieved through a gradual staging of reductions over a period of eight years.

3. For a detailed discussion of the non-tariff barriers described in this section, see U.S. International Trade Commission, Competitive Factors Influencing World Trade in Integrated Circuits, Washington, D.C., November 1979; and Organization for Economic Cooperation and Development, Telecommunications Equipment Industry Study, Paris, October 1981.

4. See, for example, "High-technology gateway: Foreigners demand a piece of NTT's $3 billion market," Business Week, 9 August 1982.

5. The impediments to successful direct investment in Japan by U.S. companies range from difficulties in recruiting able and experienced engineers to preferential access for Japanese firms to capital, government guarantees, special tax incentives, loans, and subsidies. For example, a significant barrier to entry to the Japanese market is posed by the inability of foreign companies to acquire Japanese companies. The difficulty in acquiring Japanese companies is due partly to Japanese law, which requires unanimous approval by the board of the company to be acquired, thus effectively blocking any unfriendly takeover. Cultural factors, however, also make the sale of a Japanese company unusual, except in cases of near bankruptcy or other financial difficulty. See, "Japan's aversion to selling companies may be ultimate barrier to U.S. trade," The Wall Street Journal, 23 March 1982.

6. The electronics standards certification process in the United Kingdom is also reported to constitute a barrier to imports of integrated circuits.

4

The Semiconductor Industry

Sustained innovation in semiconductor technology has been central to the strength of the U.S. electronics industry. Since the invention of the seminal semiconductor device, the transistor, over three decades ago, the state of the art in microelectronics has advanced at a phenomenal rate. New developments have provided better performance at lower costs. As each new performance/price threshold has been attained, the number of feasible applications for semiconductor technology has multiplied.

As discussed in Chapters 2 and 3, however, the future technological leadership of the U.S. semiconductor industry is threatened by increasingly severe financial and human resource constraints as well as by mounting protectionist sentiment both here and abroad. At the same time, the Japanese semiconductor industry has been successful in closing the technological gap with U.S. firms in several critical high-volume, high-growth product areas. These developments also threaten the position of U.S. firms in electronics equipment markets given the need for close interaction between systems producers and IC manufacturers in designing and producing competitive new equipment.

This chapter reviews the position and problems facing the semiconductor industry. Section one describes the size and international position of the U.S. semiconductor industry. Although international trade has steadily increased in importance to U.S. semiconductor firms over the past decade, the U.S. industry has slipped from a position of dominance to being strongly challenged. This challenge has come not just from foreign competitors, but from foreign governments that have focused on their local industries for development and offered low-cost financing, trade protection, and other measures to help develop them.

The second and third sections of the chapter examine the U.S. industry structure and the semiconductor manufacturing process. In contrast to Europe and Japan, a large portion of U.S. production is accounted for by firms with relatively low degrees of vertical integration. Easy conditions of entry in the noncaptive or merchant sector have historically promoted the technological competitiveness of the U.S. industry. In recent years, however, dramatic increases in the complexity and cost of integrated circuit manufacturing have encouraged a stream of mergers, acquisitions, and joint ventures that have altered the structure of the industry.

The final section of the chapter provides an assessment of the dynamics of U.S.-Japanese competition in random-access computer memories. Several of the factors discussed in Chapters 2 and 3--Japan's joint government/industry VLSI research program, its favorable financial environment, and its protectionist trade and investment policies--have been major ingredients of the Japanese industry's success. Japan's semiconductor industry has also been promoted by its dominant position in consumer electronics. Whether the Japanese industry is able to make further competitive inroads in the years ahead will depend on the combined response of U.S. industry, government, and universities.

INDUSTRY SIZE AND INTERNATIONAL POSITION

The U.S. semiconductor industry has historically served three distinct customer groups--consumer, industrial/commercial, and military. The computer industry today is the single largest customer of the U.S. semiconductor industry, accounting for approximately 40 percent of semiconductor production, a share that has been fairly constant for 20 years.[1] The share of semiconductor output consumed by military and aerospace applications has declined dramatically from 50 percent in 1960 to approximately 10 percent in recent years. The remaining portion, which has increased from roughly 10 percent to 50 percent today, is divided among telecommunications equipment, process control and test equipment, and consumer products, such as automobiles, calculators, television sets, and video games.

The value of U.S. product shipments of semiconductors and related devices increased from $2.4 billion to $9.1 billion between 1972 and 1981, an average growth rate of 16.2 percent.[2] Inflation-adjusted growth, however, was even higher, given price declines in important product lines.[3]

The semiconductor industry is highly dependent on international trade. In 1981, U.S. exports of semiconductors were $3.5 billion, and imports stood at $3.3 billion. Over the 1972 to 1981 period, exports increased at a 25 percent annual rate in current dollar terms, while imports went up at a 29.2 percent rate.[4]

Between 1972 and 1981, exports as a percentage of product shipments rose from 19.9 percent to 38.3 percent, while imports as a percentage of new supply (product shipments plus imports) rose from 12.3 percent to 26.5 percent.[5] Imports as a percentage of total trade increased gradually over the period from 41.2 percent to 48.4 percent, reflecting an increase in U.S. offshore operations and heightened competition from Japan, particularly in the random-access memory segment of the industry.

Because a single semiconductor device is frequently shipped between several countries before it is a completed manufacturing product, a review of the trade data can be misleading. U.S. semiconductor firms began to establish offshore operations in the 1960s in order to be more cost-competitive and to overcome foreign trade barriers to U.S. exports. Accordingly, U.S. import data must be interpreted with care since it consists not only of imports of foreign products, but also of the reimportation of completed devices originally exported as sub-assemblies. In fact, these sub-assemblies represent as much as 75 percent of the value of imported integrated circuits.[6]

The U.S. semiconductor industry has also faced increasing competition from the Japanese semiconductor industry. Although the domestic Japanese semiconductor market has grown dramatically over the past decade and is now the second largest in the world, U.S. exports to Japan face formidable trade barriers. Since 1977 the United States has had a deficit in total integrated circuit trade with Japan.[7] In 1978 the United States ran a $3.7 million trade deficit with Japan in integrated circuits. By 1980 that deficit had reached nearly $300 million. Even in advanced Metal-Oxide Semiconductor (MOS) integrated circuits, where U.S. firms have been technology leaders, U.S. exports to Japan increased by only $25 million between 1975 and 1980. Conversely, Japanese exports of advanced MOS ICs to the United States went from practically nothing in 1975 to $120 million in 1980.

In contrast to the U.S. semiconductor industry's lack of success in Japan, U.S. firms currently supply half of Europe's total semiconductor demand and more than 60 percent of its integrated circuit consumption. While the protectionist policies of the Japanese government effectively prohibited the establishment of U.S. manufacturing subsidiaries until recently, U.S. semiconductor firms made significant investments in Europe beginning in the late 1960s. As discussed in Chapter 3, this movement was influenced by high European tariffs as well as by pressure from European governments to establish local production facilities. In some cases, establishing a manufacturing facility was a prerequisite to obtaining access to segments of particular national markets, e.g., communications in the United Kingdom and computers in France. Currently, four U.S. firms, all with major European production facilities, are among the 10 largest European semiconductor producers.[8]

U.S. semiconductor firms, however, are likely to face increased competition in Europe over the next decade. Japanese electronics firms have recently moved to establish semiconductor production facilities in Europe. Moreover, in addition to the historical protectionist barriers in western Europe, national governments, particularly those in France, West Germany, and the United Kingdom, have moved to assist their local industries in developing capabilities in advanced microelectronics, e.g., through direct governmental support, preferential purchasing arrangements, and acquisition of U.S. technology.

INDUSTRY STRUCTURE

The U.S. semiconductor industry has moved through three phases in its 30-year history. The first phase, which started with the development of the transistor and continued through the 1950s, saw the evolution of basic semiconductor technology. This evolution permitted greater mechanization of the transistor manufacturing process, improved performance and reliability, increased production rates, and lowered unit costs, resulting in turn in lower prices and greater sales volume. The second phase was ushered in with the introduction of the integrated circuit. During this phase, which lasted through the early 1970s, the number of semiconductor elements on a single IC chip increased steadily through medium-scale integration to the early stages of large-scale integration, where as many as 10,000 component elements were fabricated on a single chip. The third phase, which began in the early 1970s, has been characterized by advances in large-scale and very large-scale integration, notably the incorporation of logic and memory on the same chip, that have yielded dramatic improvements in the cost and performance of the latest electronic equipment.

The initial phase of the semiconductor industry's development began with the establishment of semiconductor divisions by major electronic tube manufacturers. Production in these new facilities was for either internal use (captive) or for the open market (merchant). Western Electric and IBM established captive facilities--a result of court decree in the first case and corporate choice informed by antitrust considerations in the latter. Large tube and equipment manufacturers such as RCA, General Electric, and Sylvania established semiconductor operations to supply both their internal needs and the open market. In addition, Texas Instruments and Motorola, though not tube manufacturers, followed a similar pattern and are now major merchant producers.

Three important factors encouraged the development of the semiconductor industry in the 1950s. The first was the U.S. military's interest in the technology.[9] For a limited time, the U.S. military was a major sponsor of the development of new devices. Further, the military's willingness to pay a premium for quality and reliability helped the industry to bear the cost of refining and debugging its products. Second, the establishment of liberal technology licensing policies broadened the industry's technological base and encouraged its expansion.[10]

The introduction of the integrated circuit in the early 1960s widened the reach of semiconductor technology, enabling the industry to reduce its unit costs and thereby its dependence on military demand. The growth of the computer industry and its price-sensitive demand for ICs created opportunities for entry by new producers. Entry of new firms was also encouraged by the availability and cost of capital, the availability of labor, and the technological capability of the individuals forming new companies. The 1960s thus witnessed an influx of new entrants, including some of today's leading merchant producers such as Advanced Micro Devices, Intel, Mostek, and National Semiconductor. An important spawning ground for new ventures during the 1960s was Fairchild Instrument and Camera, the innovator of the planar process that made integrated circuitry commercially feasible. Fairchild's Semiconductor Division itself had only been established in 1957.

The new entry phenomenon that characterized the semiconductor industry in the 1960s was critical in stimulating competition in product design, price, service, and quality. Further, because the advent of IC technology opened new opportunities for product differentiation, new entrants were able to exploit market niches overlooked or underdeveloped by established firms. As shown in Table 4-1, new entry and the dynamics of competition led to a marked decline in the concentration of IC shipments between 1965 and 1978.

After the early 1970s, however, the rate of new entry into the semiconductor industry slowed dramatically, while the number of mergers and acquisitions increased.[11] These trends are attributable to three factors. First, the capital equipment and R&D expenditures necessary to keep up with the latest technology have increased the cost of entry. Rising capital costs, coupled with changes in the late 1960s in the tax treatment of capital gains, increased the difficulty of forming new ventures. Second, the evolution of semiconductor technology has encouraged systems manufacturers to integrate backwards. As individual ICs have taken on the capability of complete electronic systems, systems producers have sought to acquire merchant semiconductor firms as a means of incorporating proprietary designs, assuring a reliable source of supply, and capturing the increasing value added

TABLE 4-1   Concentration of U.S. Integrated Circuit Shipments

Number of Companies	Percent of Total Shipments			
	1965	1972	1978	1982
4 largest companies	69	53	49	47
8 largest companies	91	67	70	64
20 largest companies	99	94	90	84
50 largest companies	100	100	100	100

SOURCE: The Semiconductor Industry Association, *The International Microelectronic Challenge*, Cupertino, CA, May 1981, and Dataquest.

associated with the design and production of large-scale integration devices. Finally, acquisitions by foreign electronics firms were encouraged by the weakness of the U.S. dollar in the 1970s and, until recently, by the depressed state of the U.S. stock market. These factors made it cheaper to buy technology (and U.S. market share) through the purchase of existing companies than to gain it through in-house development or adaptation of the latest technologies.

Recently, however, the rate of entry into the semiconductor industry has increased with an influx of new companies specializing in the design or production of custom integrated circuits.[12] This wave of entry has been encouraged both by the increased availability of venture capital following the 1978 revision of the capital-gains tax law and by the increased demand for custom circuits generated by the move to VLSI technology. Existing companies have also formed custom divisions, spurred by the need for closer interaction between systems manufacturers and IC producers created by the increased costs--and opportunities--of VLSI technology.[13] In some cases, the design work is performed by the customer with the IC manufacturer providing only a device processing or foundry service.

The size of the independent merchant sector of the U.S. semiconductor industry is unique compared to Europe and Japan. The major foreign producers of ICs (other than U.S. subsidiaries) are generally part of vertically integrated electronic equipment companies. Leading European producers are the Netherlands-based consumer electronics firm, Philips, and the West German-based industrial electronics firm, Siemens. Similarly, the major Japanese IC producers--Nippon Electric, Hitachi, Toshiba, Mitsubishi, Matsushita, and Fujitsu--are also manufacturers of electronic systems.

The vertically integrated structure of foreign semiconductor industries is an advantage to the extent that it facilitates close interaction between component and systems divisions and makes capital available from internal funds outside the IC division. Nevertheless, the size and sophistication of the U.S. merchant sector offers the industry as a whole important strengths that foreign industries lack. These include the merchants' ability to achieve low-cost volume production by standardizing designs for many customers' needs, the role of merchant producers in accelerating the pace of technological advance,[14] the ability of captive companies to rely on merchant supply for peaks in their demand,[15] and the promotion of a strong IC production equipment industry. It is thus difficult to view either U.S. merchant or captive producers in isolation. Rather, the health of one depends on the other, and the strength of the U.S. electronics industry depends on the two together. Indeed, increasing recognition of the interdependence between component suppliers and systems manufacturers is reflected in a wave of cooperative initiatives--major joint ventures, technology exchanges, and ownership agreements--among computer, communications, and semiconductor firms.[16]

THE SEMICONDUCTOR MANUFACTURING PROCESS

The manufacture of semiconductor devices requires mastery of highly complex production techniques. What makes the process economic is the use of mass production techniques. Basically, semiconductor manufacturing is a batch process with a large number of precise operations involving heat treatment and multiple stages of chemical deposition and etching.[17]

Circuit designs are first transferred to a "mask," which is used as a template in the device fabrication process. A number of identical integrated circuits are then created on a single circular wafer of silicon. Even in the most successful wafer fabrication, some of the integrated circuits will have defects. Problems such as impurities may prevent a particular integrated circuit from functioning properly. It is therefore necessary to conduct extensive testing of the fabricated circuits at each stage. One round of testing is conducted before the wafer is cut into individual chips, and the defective chips are marked. After cutting, the defective circuits are discarded and the good ones sent on for bonding into packages that protect the integrated circuits and provide electrode connections for insertion into printed circuit boards. Another round of testing is conducted after the assembly into packages.

Because of the complexity of the production process, only a percentage of the devices out of the total number fabricated are actually acceptable at the end of the manufacturing sequence. The ratio of successful devices to the batch total is referred to as the "yield." Yield is a measure of the efficiency of the production process. For the first commercial runs of the most sophisticated types of integrated circuits, the initial yields have historically been low. However, over time, improvements in yield are made as production experience is cumulated. A recently introduced method of increasing yield for some high-density ICs is to build in redundant components and use laser beams to make new circuits when certain portions of the chip are found to be defective. As a result of such process improvements, yields for mature products typically become quite high. This phenomenon--of increasing yields over the life of a product--is one of the basic underpinnings of the learning curve in semiconductor manufacturing.

The strength of the learning curve in this industry provides a strong incentive to either be the manufacturer with the first widely accepted design, or if not the originator, to build volume rapidly with a lower priced copy or variation of the widely accepted design. By increasing volume, a manufacturer reduces its costs and either increases its profit or is able to cut prices and build further volume. The only way to escape the pressure of other firms moving down the learning curve, in essence, is to create a new learning curve by making a product advance sufficiently great to obsolete much of the previous product.

Another aspect of IC production that puts downward pressure on price is the ease with which devices can be disassembled, analyzed, and functionally replicated. Although some protection is available through patents and copyrights, the litigation process is time-consuming compared to the pace of competition. In any case, patent protection is relatively weak, as it is in other areas of electronics, due in part to the high degree of interdependency among different inventions and components of a system. Therefore, there is considerable cross-licensing of IC patents and willingness to allow other firms to produce or second source a particular design. This willingness is reinforced by purchaser requirements for second-source protection. Since the cross-licensing is well institutionalized--due in part to AT&T's adoption of an open-licensing policy in the 1950s--the major competitors need not necessarily actually disassemble another firm's device to copy it, but merely obtain a license to get the design. Typically, they would then not produce an exact copy, but rather a compatible device, with their own improvements added. Nevertheless, the potential for copying does result in some actual instances of that practice and also contributes to the willingness of firms to license widely.

Finally, as discussed in Chapter 2, the advent of large-scale and very large-scale integration technology has substantially increased the capital investment required to manufacture advanced integrated circuits. These technologies have required

new production equipment, whereas prior semiconductor generations could be produced by equipment already in place. For example, the 16K RAM could be produced with basically the same type
of equipment that had been used to produce the predecessor
device, the 4K RAM; in contrast, the equipment necessary to go
from the 16K RAM to the 64K and beyond is different and more
costly. Similarly, equipment used in other phases of production is
rapidly escalating in cost. Testing in particular has become more
costly as product quality competition has intensified, and the
devices themselves have become more complex and difficult to
test. Available estimates indicate that testing costs have increased from about 10 percent of total cost to 25 percent.[18]

BASES FOR COMPETITION:
A CASE STUDY OF THE RANDOM-ACCESS MEMORY

High-volume products, such as MOS memory devices for computers, have great strategic importance to the international
competitive position of the U.S. industry. Profits on high-volume
products help finance the research required for new products.
Further, because MOS memory devices are one of the largest
semiconductor product groups in terms of sales volume, these
products stand at the cutting edge of IC technology. It is an area
in which production techniques have been advanced and refined
and, therefore, have led the way technologically to greater IC
density and complexity. A company that is not involved in MOS
memory technology risks its position in the semiconductor
industry, even as a specialized supplier.[19]

Until the late 1970s, the U.S. semiconductor industry was
unchallenged in IC technology and markets. Because the U.S.
computer industry was the largest in the world, and because
computers used the most advanced ICs, it was difficult to imagine
how the dominant position of the U.S. semiconductor industry
might ever be seriously threatened. In contrast, the Japanese had
concentrated on integrated circuits for consumer electronics and
had little capability for circuits used in computer and telecommunications equipment. In 1976, however, as part of a general
effort to develop those industries, the Japanese launched their
four-year, joint government/industry VLSI program to produce
more complex integrated circuits.

The dynamic Random-Access Memory, a high-volume, standardized product, was an ideal first target for the Japanese. It
allowed them to capitalize on their traditional strength (low-cost
manufacturing) and to minimize their relative weaknesses (innovation and distribution/support services). In addition, capital and
technical personnel problems hampered the competitive reactions

of U.S. semiconductor firms, aiding the Japanese assault, first in 16K and later in 64K RAMs.

The ability of the Japanese to compete successfully in state of the art computer memory devices was dramatically demonstrated in the 16K RAM.[20] When introduced in late 1976, the 16K RAM was considered the most advanced semiconductor memory device in volume production. Three U.S. companies began shipping the 16K RAM to customers in late 1976. In 1977 they were joined by two other U.S. and two Japanese companies. The Japanese followed an entry strategy based on second sourcing the 16K design of Mostek, the U.S. market leader.

A complex interaction of market factors characterized the 1978 to 1980 period. Demand was exceptionally strong due to the strength of the U.S. economy and an unanticipated requirement by IBM and others for a substantial quantity of 16K RAMs. During the 1974 to 1975 recession, however, U.S. semiconductor firms had cut back sharply on capital spending for production capacity. By 1979 demand for the 16K RAMs picked up to the point where it overwhelmed the capacity-limited output of U.S. manufacturers.

In addition, the demand for the 16K RAM was very price-sensitive because, like other high-volume semiconductor products, the device was characterized by standard specifications. To improve their market position, Japanese suppliers repeatedly lowered their prices. In late 1977 prices of the 16K RAMs ranged from $16 to $18; by late 1978 prices had fallen to the $5 to $6 range. This price pressure led several U.S. firms to discontinue RAM production. Moreover, early U.S. buyers of Japanese 16K RAMs, notably Hewlett-Packard, reported that Japanese RAMs had a lower frequency of defects than domestic ones. Thanks to aggressive automation, the Japanese thus took the lead in setting 16K RAM yield, quality, and reliability standards. By the end of 1979 the Japanese had garnered over 40 percent of U.S. sales of 16K RAMs.

In anticipation of the 64K RAM, U.S. semiconductor manufacturers moved to neutralize the Japanese advantages in production capacity and quality. Moreover, most U.S. merchant firms pursued design strategies that sought to minimize the size of the die on a 64K RAM, since die size potentially allows the manufacturer to increase the number of usable chips obtained from each wafer.[21] In contrast, the Japanese chose a conservative design approach, adapting and improving a design that had been used in Mostek's 16K RAM. As a consequence of designing larger cells on larger dice, the Japanese were more successful than their U.S. counterparts in avoiding technical problems and establishing volume production. By the end of 1981, the Japanese held an estimated 70 percent of 64K RAM sales, and only two U.S. firms, Motorola and Texas Instruments, had begun volume production.[22]

Most observers, however, expect U.S. semiconductor manufacturers to improve their product designs and production efficiency and, therefore, to increase their market share as sales of 64K RAMs expand.[23] Until recently, users had little incentive to replace 16K RAMs in their products because recession and competition combined to cause a dramatic collapse in the prices of the older generation products. However, as the price of 64K RAMS has also plummeted, demand has exploded.[24] By 1983 worldwide sales are forecast to be nearly $1 billion--a figure that would make the 64K the largest selling product in semiconductor history. The booming demand for 64K RAMs, by creating tight supplies and stretched deliveries among some suppliers, has thus provided a market opening for late U.S. entrants.[25]

The competitive position of U.S. suppliers may also be aided by U.S. users in the computer industry. Reluctant to rely completely on Japanese suppliers such as Hitachi, Fujitsu, and NEC, which are also competitors, U.S. systems manufacturers have reportedly waited to qualify U.S. firms on their list of approved 64K RAM suppliers, even though the qualification process involves substantial costs. In addition, U.S. semiconductor firms retain the advantage of being broad-based suppliers, while Japanese product offerings are more limited.

Nevertheless, with their success in 64K RAMs, Japanese semiconductor manufacturers have enhanced their competitiveness in successor generations.[26] By gaining early experience in making the complex 64K RAM, the Japanese are widely expected to capture a substantial share of 256K sales. In early 1983, at least four of the top Japanese semiconductor manufacturers--Hitachi, NEC, Fujitsu, and Toshiba--were reported to be sending samples of their 256K RAMs to potential customers. Indeed, industry observers believed that some Japanese producers were also prepared to begin mass production and mass marketing.[27] In contrast, not one of the traditional U.S. memory manufacturers was close to starting volume production.

Problems in the three major areas discussed in Chapters 2 and 3--availability of technical personnel, cost of capital, and restricted access to foreign markets--have had a major impact on the competitiveness of U.S. RAM manufacturers vis à vis the Japanese. High staff turnover resulting from shortages of key technical personnel contributed to the difficulty experienced by U.S. firms in achieving their sophisticated 64K design objectives. The 64K effort has required longer range planning and greater team work than previous products. Accordingly, problems in keeping design teams together have hindered the efforts of U.S. semiconductor firms, particularly those located in Silicon Valley.[28]

Personnel continuity is also important on the production side. The need for training and retraining of workers has placed a strain on U.S. semiconductor manufacturers that is minimized under the lifetime employment systems of large Japanese electronics firms. Indeed, the greater continuity of Japanese workers in their jobs is frequently cited as an important factor in explaining the high quality of Japanese ICs.

At this time, however, the most pressing problem for U.S. firms arises from Japanese capital-cost advantages. For example, U.S. semiconductor firms increased investment spending slowly following the sharp cutback made during the 1975 recession. In contrast, investment spending by Japanese firms accelerated in 1976, spurred by a boom in consumer electronics sales. Japanese firms were, therefore, able to take advantage of the rapid growth in demand for 16K RAMs in the late 1970s. The Japanese semiconductor industry is again reported to be engaged in a major investment campaign.[29] This investment surge comes at a time when U.S. semiconductor manufacturers, suffering from weak markets, high capital costs, and aggressive Japanese competition, have had to reduce operations and delay many investment programs.

Restricted access to the Japanese market has also contributed to the Japanese success in winning RAM market share. According to the Semiconductor Industry Association, Japanese prices for 16K RAMs in the United States were some 20 to 30 percent below levels in Japan in 1979. Such price discrimination is difficult to achieve without a sheltered market. The higher prices in Japan thus contributed to R&D and capital investment for the 64K generation, while the lower price in the United States succeeded in denying revenue to U.S. firms for future product development and production capacity. (The problem of restricted access to the Japanese market is, of course, compounded by the vertical ties between Japanese IC producers and systems equipment manufacturers.) Although the Japanese have eased restrictions on IC direct investment considerably since 1979 (as was done for consumer electronics in the early 1970s), the restrictions were not lifted until the Japanese industry was established as a formidable competitor.[30]

The Japanese success in 64K RAMs also raises fundamental questions about the ability of U.S. firms to counteract Japanese competition in other segments of the IC industry. In microprocessors, the Japanese lag U.S. manufacturers in circuit design skills and software technology. The Japanese leaders--NEC, Hitachi, and Matsushita--so far have only been able to capture market share in relatively low-value, low-power microprocessors. In 4-bit microcomputers, used mainly in consumer electronics products that require relatively little in the way of software support, large volume has enabled the Japanese to capitalize on their highly automated production techniques and their large domestic market for consumer-oriented components. In the more complex 8-bit and 16-bit microprocessors, which go into computers and office machines and require larger amounts of software, Japanese manufacturers are currently either copying or licensing U.S. designs.[31] Nevertheless, MITI is currently coordinating a joint software development program that should not be underestimated in its ability to give the Japanese industry a competitive boost.

Although the Japanese are currently weak in microprocessors and other logic circuits, Japanese semiconductor manufacturers must diversify into these areas or their memory business may suffer to the extent that customers prefer dealing with broad-line suppliers. Moreover, some observers believe that Japanese firms will begin to develop more complex microprocessors when a large volume export market in personal computers materializes. Indeed, the Japanese lag is expected to narrow as standardized software and system designs are adopted for microprocessor applications, particularly in personal computers.

Finally, an important factor supporting the Japanese drive to develop their competence in integrated circuit technology is the countercyclical support afforded by the strength of the consumer electronics industry in Japan. Indeed, consumer electronics has been the driving force behind the development of the Japanese semiconductor industry. In 1981, for example, Japanese consumer electronics, led by worldwide demand for home video cassette recorders, accounted for 50 percent of Japan's semiconductor output.[32] By contrast, just 15 to 20 percent of U.S. and European semiconductor production goes into consumer goods. Consumer electronics sales have thus bolstered Japanese semiconductor production during periods when the more heavily computer-dependent U.S. industry has suffered slowdowns.

NOTES

1. Robert W. Wilson, Peter K. Ashton, and Thomas P. Egan, Innovation, Competition and Government Policy in the Semiconductor Industry, A Charles River Associates Research Study. D. C. Heath & Co., Lexington, Massachusetts, 1980.

2. In contrast to the 1972 to 1981 period as a whole, the value of product shipments in 1981 increased by only 2.0 percent in current dollars (2.5 percent in 1972 dollars). U.S. Department of Commerce, U.S. Industrial Outlook 1982, Government Printing Office, Washington, D.C., 1982.

3. The U.S. Department of Commerce reports an average inflation-adjusted growth rate of 19.3 percent for the 1972 to 1981 period. However, the government price index seriously understates the rate of price decline and, therefore, the inflation-adjusted growth rate, due to the rapid rate of product change, the sharp price declines in new products relative to older products, and the lag in incorporating new products into the government price index.

4. The increases in U.S. exports and imports in 1981, 1.5 percent and 4.3 percent, respectively, also stand in contrast to the 1972 to 1981 period as a whole. U.S. Department of Commerce, U.S. Industrial Outlook 1982, Government Printing Office, Washington, D.C., 1982.

5. U.S. Department of Commerce, U.S. Industrial Outlook 1982, Government Printing Office, Washington, D.C., 1982.

6. In 1978 imports of semiconductors under the special tariff provisions of items 806.30 and 807.00, which enable U.S. manufacturers to reimport products paying duty only on the value added by their offshore facilities, amounted to $1.5 billion. U.S. International Trade Commission, Competitive Factors Influencing World Trade in Integrated Circuits, November 1979.

7. U.S. Department of Commerce, U.S. Industrial Outlook 1982, Government Printing Office, Washington, D.C., 1982; and The International Microelectronic Challenge, The Semiconductor Industry Association, Cupertino, California, May 1981. The U.S. Department of Commerce may underestimate U.S. exports to Japan to some degree because U.S. firms export to Japan from their offshore facilities as well as directly from the United States. Nevertheless, these measures of the trade balance in semiconductors do not include semiconductor components in electronic equipment imported into the United States from Japan. In 1978, for example, the value of the semiconductor content contained in electronic equipment exported by Japan to the United States has been estimated to be almost three times the value of visible component exports. World Semiconductor Industry in Transition 1978-1983, Arthur D. Little, Cambridge, Massachusetts, 1980.

8. The International Microelectronic Challenge, The Semiconductor Industry Association, Cupertino, California, May 1981.

9. The earliest ICs were based on bipolar semiconductor technology, which possesses properties that are well suited to military applications, especially for equipment operating under harsh environmental conditions. Metal-Oxide Semiconductor (MOS) ICs, which appeared in the mid-1960s, operate at lower speeds than their bipolar counterparts, but consume less power. Lower power consumption permits a denser packing of circuit

elements since less heat has to be dissipated. As a result of this density advantage, MOS devices have gained primacy in commercial computer applications. Although bipolar technology remains important in many applications due to its performance advantages, advanced complementary MOS (C-MOS) technology of small dimensions is expected to surpass bipolar in terms of speed.

10. The major U.S. technological innovators in the semiconductor industry have followed a liberal licensing policy, which recognizes the difficulty of maintaining industrial secrecy in an industrial environment where mobility of scientists and engineers has been high and results of semiconductor research were widely available. For a discussion of technology licensing during the early years of the industry, see John E. Tilton, International Diffusion of Technology: The Case of Semiconductors, The Brookings Institution, Washington, D.C., 1971.

11. The net effect of the acquisitions and mergers over the past decade on the ownership interests of U.S. merchant IC production has been dramatic. In 1970 there was no foreign ownership of any significance; by 1978 nearly one-fifth of total U.S. merchant IC production was concentrated in firms with significant foreign ownership. Likewise, the proportion of production under the control of large parents (excluding Texas Instruments and Motorola) went from a negligible amount to 11 percent by 1978. The International Microelectronic Challenge, The Semiconductor Industry Association, Cupertino, California, May 1981. Merchant firms acquired since 1978 include Synertek (Honeywell), Fairchild Camera & Instrument (Schlumberger), Mostek (United Technologies), Siliconix (Westinghouse), and Intersil (General Electric).

12. See, for example, "New starters in Silicon Valley," Business Week, 26 January 1981.

13. In addition to the activity in custom devices, existing merchant manufacturers, e.g., Intel, National, TI, and Motorola, are attempting to integrate forward into office automation equipment. See, for example, "Intel may soon compete with its customers," Business Week, 22 March 1982, and "Motorola's new strategy: Adding computers to its base in electronics," Business Week, 29 March 1982.

14. On the subject of technological change, one study found that over 80 percent of the major semiconductor innovations between 1960 and 1977 were introduced by merchant companies. (See Wilson, et al., op. cit.) Although some of these innovations were built on research advances made at captive companies, the merchant sector has been effective at accelerating the process of technological change. The captive producers, in turn, benefit both from having the technology advance further and from the availability of advanced devices from outside sources.

15. Most captive IC production is run near capacity, with peak needs of IC devices bought from merchant producers. Thus, the merchant companies tend to suffer sharper cyclical fluctuation in demand than captive producers.

16. For a listing of recent cooperative agreements, see "IBM and Intel link up to fend off the Japanese," Business Week, 10 January 1983.

17. For a more detailed description of the semiconductor manufacturing process, see U.S. Department of Commerce, A Report on the U.S. Semiconductor Industry, Government Printing Office, Washington, D.C., 1979.

18. The International Microelectronic Challenge, The Semiconductor Industry Association, Cupertino, California, May 1981.

19. Jerry Sanders, president of Advanced Micro Devices, has explained the importance of computer memories as follows: "The way to drive down costs for almost any integrated circuit is to make computer memories. They are the single part with the highest volume, and the process art you acquire producing them can be applied to everything else. By making complex parts, you also learn how to make your simpler consumer circuits perform better." As quoted in Bro Uttal, "Europe's wild swing at the silicon giants," Fortune, 28 July 1980.

20. This account of Japanese competition in 16K and 64K RAMs draws on the following sources: "Japan's ominous chip victory," Fortune, 14 December 1981; "Japan's strategy for the '80's," Business Week, 14 December 1981; "Rolling with recession in semiconductors," Business Week, 21 July 1980; "Can semiconductors survive big business?, Business Week, 3 December 1979; "Japan's big lead in memory chips," The New York Times, 28 February 1982; "The chip makers' glamorous new generation," Business Week, 6 October 1980; "A chance for U.S. memories," Business Week, 15 March 1982; "Two chip-making giants gear up for recovery," Business Week, 31 May 1982.

21. The size of the die in a chip, however, is not of great significance to users since packaged chips of a particular type are all the same size.

22. Motorola, like the Japanese companies, followed a conservative design strategy. TI, which adopted a more complex design, was later than Motorola in establishing volume production.

23. For example, IBM selected Intel to provide second-generation 64K RAM design and process technology. Although IBM already makes its own 64K RAM, Intel's new 64K RAM offers superior speed, redundancy, and a sophisticated process technology known as "direct-step-to-the-wafer." "Intel wins memory-chip job from IBM, a major coup in Japan-dominated field," The Wall Street Journal, 2 September 1982.

24. To solidify their market position, the Japanese have repeatedly lowered prices. As recently as March 1981, prices for 64K RAMs were $28; by March 1982, prices had dropped to the $6 to $10 range, with further price reductions anticipated. Indeed, some observers believe that, at such levels, Japanese prices were below the cost of production.

25. To be profitable at the current low price levels, new 64K suppliers, as well as other suppliers, are attempting to redesign their products to reduce costs and improve product reliability. Only firms with the lowest manufacturing costs will be able to compete for sales of 64K RAMs.

26. The Japanese will enjoy a significant advantage if their designs become de facto product standards. To the extent that other firms incur higher costs as a result, the standard setter will be able to exploit its cost advantage to expand market share and to finance future product development.

27. Although the major Japanese semiconductor manufacturers are pushing ahead with development of the 256K dynamic RAM, they face a strategic dilemma. If they move too aggressively into the 256K, sales of 64Ks will collapse. Nevertheless, competition among the Japanese companies may lead them to mass produce the 256Ks sooner than they might have hoped.

28. The ease in obtaining venture capital is exacerbating the turnover problem by prompting increasing numbers of key employees from established manufacturers to leave and form new companies. See, for example, "Technology: Venture capitalists raid Silicon Valley," Business Week, 24 August 1981.

29. See, "Capital investment in semiconductors expanded by Japan," The New York Times, 1 April 1982. The Japanese semiconductor industry is guiding its investment plans to guard against protective action by foreign governments. For example, the Japanese are increasingly investing in semiconductor production and assembly facilities in the United States.

30. Until the liberalization measures, the Japanese effectively prohibited both the construction of foreign-owned plants in Japan and foreign investment in existing Japanese semiconductor companies. The sole exception, Texas Instruments, was able to establish a manufacturing presence in Japan in the late 1960s due to its ability to withhold key IC production patents from the Japanese unless access was granted. Without these patent rights, exports of Japanese products using ICs could have been challenged.

31. However, in October 1981 NTT announced the development of a 32-bit microprocessor, only months after Intel, Bell Laboratories, and Hewlett-Packard announced their own designs.

32. For example, in the first half of 1981 MITI estimated that Japanese semiconductor production was up 24 percent, while U.S. production for 1981 as a whole was down 2 percent. See, "Semiconductors: Consumer electronics provides the foundation," Business Week, 14 December 1981.

5
The Computer Industry

The three-decade history of the U.S. computer industry has been one of extremely rapid technological change. Order of magnitude increases in computing power have been accompanied by order of magnitude decreases in cost per calculation. Computers have been used on a far broader and deeper scale than predicted by government or business in the early years of the industry. U.S. firms have been leaders in computer product and application development throughout the post-World War II period.

This chapter examines the competitive position of the U.S. computer industry. The first section describes the overall size and international position of the U.S. industry. U.S. firms account for
an impressive share of both domestic and international computer sales, though Japanese firms have captured large market shares in some low-cost peripheral equipment products. Nevertheless, the efforts of our trading partners, particularly Japan, are focused directly on taking the competitive lead from the United States.

The second and third sections of the chapter present an overview of the history of the industry and the underlying technological developments in semiconductor logic and memory. New entrants have had an important influence on the structure of the computer industry, particularly in pioneering new product segments. Changes in the industry's structure have also been closely related to the evolution of semiconductor technology. For example, the entry of new firms producing personal computers was made possible by developments in integrated circuit microprocessors.

The final sections examine the bases for competition in the computer industry, with particular emphasis on the relative U.S. position in software development, production capabilities, and distribution networks. The nature of competition in the industry is changing as the economics of volume production have turned products such as computer memories and microcomputers essentially into commodities, thereby heightening the importance of customer-tailored software and service. The U.S. strength in software development and distribution will thus continue to be an important competitive advantage. Nevertheless, the ability of the Japanese to close the software gap must not be underestimated.

INDUSTRY SIZE AND INTERNATIONAL POSITION

The computer industry has been one of the most remarkable growth industries of all time. The value of computers and related equipment shipped in the United States was just over $1 billion in 1958, growing to $4.2 billion in 1968, and $16.6 billion in 1978. Shipments in 1981 were estimated by the U.S. Department of Commerce at over $30 billion, giving the industry a compound annual rate of growth of 18.8 percent between 1972 and 1981. An explosion of the customer base is expected in the 1980s. One forecast predicts that the value of industry shipments will reach $74.8 billion by 1985.[1]

Exports of computer equipment and parts, an important item in the U.S. balance of payments, rose 20 percent in 1981 to an estimated $8.8 billion.[2] This increase was less than the 30 percent annual rate of growth enjoyed between 1976 and 1980 due to the adverse effects on U.S. computer exports of the overvaluation of the U.S. dollar in relation to the Japanese yen and downturns in key western European markets. In addition, the value of exports showed a shift toward a greater percentage of parts over equipment. For example, Ireland and Hong Kong have recently emerged as important importers of U.S. computer equipment due primarily to the buildup in these countries of U.S. assembly facilities, particularly those established by several leading U.S. minicomputer and microcomputer firms. Imports of computing equipment and parts grew rapidly in 1981, rising an estimated 38 percent to $1.6 billion.[3] As a result, the ratio of imports to new supply (product shipments plus imports) increased from 0.045 to 0.052. Canada and Japan remained the leading suppliers, while imports from the United Kingdom and France fell below their 1980 levels. In the case of Japanese exports to the United States, joint marketing ventures between Japanese suppliers and U.S. distributors helped to boost the 1981 level. Finally, parts assumed a larger share of imports from many European countries and represented more than 90 percent of import value from Hong Kong and Mexico. U.S. subsidiaries in many of these countries were significant contributors to these shipments.

The competitive strength of the U.S. computer industry, both domestically and internationally, is highlighted by comparison of

TABLE 5-1   1980 Trade Ratios for Computer Equipment of Principal
Nations

Country	Exports/ New Supply	Imports/ New Supply
United States	0.293	0.045
France	0.225	0.267
West Germany	0.364	0.423
Japan	0.121	0.156
United Kingdom	0.426	0.492
Italy	0.348	0.519

SOURCE: Official government statistical publications of each country. As presented in
the U.S. Department of Commerce, *U.S. Industrial Outlook 1982*, Washington, D.C.:
U.S. Government Printing Office.

1980 trade ratios for six principal nations shown in Table 5-1.
While West Germany, the United Kingdom, and Italy exceeded the
U.S. export-to-new-supply ratio, the United States had the lowest
import-to-new-supply ratio of the six, reflecting the U.S. industry's dominant hold on its domestic market. Japan's ratios show
the second strongest domestic position, but relatively low international sales.

Table 5-1 does not portray the full strength of the U.S. computer industry because overseas subsidiaries of U.S. computer
manufacturers contribute significantly to the computer production and trade of most of these countries. For example, in the
case of computer exports from the United Kingdom, as much as
70 to 80 percent is estimated to come from U.S.-owned operations
there.[4] While U.S. production accounted for 40 percent of the
$12.8 billion in exports of the six principal trading countries in
1979, the U.S. Department of Commerce estimates that the U.S.-
owned share would be 60 to 70 percent, or about $8 billion.

This strong U.S. position is also reflected in company-level
comparisons of the leading computer firms in these six countries.
In 1980 the worldwide computer revenues of the top eight U.S.
computer firms reached an estimated $43.7 billion, a 15.7 percent
increase over 1979.[5] Comparable revenues from eight foreign
counterparts in 1980 were an estimated $10.0 billion, up 17.1
percent over the previous year. The relative commitment of
these U.S. and foreign firms to computer products can be seen in
the fact that in 1979 the computer-related revenues as a percent
of total revenues represented 73 percent and 21 percent for the
U.S. and foreign firms, respectively.[6]

The leadership of the United States in computers, coupled with
the realization that this industry is an increasingly critical sector
of a nation's economy, has motivated both developed and developing countries alike to create or strengthen their own industries. In

several cases, governments have encouraged joint ventures, licensing, and acquisitions as a means of rapidly acquiring the necessary technology. Recently, as the technical proficiency of Japanese firms has advanced, they have become alternative sources to U.S. computer firms for new technology. Japanese companies, for example, are principal participants in joint ventures with indigenous firms in the Brazilian government's efforts to create a minicomputer industry.

Japanese computer manufacturers have become increasingly aggressive in the smaller OECD countries and in developing country markets, where they have reportedly offered deep price discounts to buy market share. For example, in Spain, Australia, and Brazil the Japanese have provided substantial discounts to central banks and government agencies to replace IBM computers with Japanese plug-compatible machines, which can run on IBM software, but offer more processing power at a lower price. Once the public sector replaced its IBM computers with Japanese models, the private sector followed, though at higher prices.[7]

Japanese computer companies are also forming alliances with several European computer and business equipment firms for the marketing of their large-scale to very large-scale computer processors. These cooperative efforts have given Japanese companies market access without the expense of establishing distribution and maintenance networks and allowed the European firms to avoid the high costs of developing production capability in this market sector.

INDUSTRY STRUCTURE

The U.S. computer industry has historically been an international technology leader. During the 1940s, laboratory work in computer technology was undertaken in both the United States and Europe; however, the U.S. industry quickly progressed from theoretical advances to develop computer products to sell in the marketplace. In 1951 a U.S. company, Remington Rand, produced the world's first commercially available, large-scale electronic computer, the Univac I. Other first-generation U.S. computers were the Univac II and IBM 701 and 702. During the 1950s, large U.S. corporations acquired first-generation vacuum tube computers to handle accounting and payroll functions. The federal government was another important customer. Indeed, the first Univac I was installed at the U.S. Bureau of the Census.

In the late 1950s, transistorized second-generation computers appeared on the market in the United States. These included the IBM 7090 and 1401, RCA 300, Control Data Corporation CDC

1604, Burroughs 3500, and Univac 1108 II. Second-generation computers were characterized by their use of transistors, advances in logic, and introduction of magnetic core memories. Data entry was on punched cards, and output was through line printers. Systems software and compilers came into use. These computers, which operated in a batch mode, were utilized in a broader range of applications than first-generation machines.

By 1960 six U.S. mainframe manufacturers--Burroughs, Control Data, Honeywell, IBM, National Cash Register (NCR), and Sperry Univac--had all entered the industry, and IBM had emerged as the industry's dominant firm. Two other entrants in the early period, General Electric and RCA, later withdrew from the industry. Xerox Corporation withdrew from the industry in 1975.

Third-generation computers were introduced in the mid-1960s. Technological advances in electronic components were central in the development of this generation. These computers incorporated integrated circuit technology, larger and faster memories, modularity in design, and time-sharing capability--features that reduced computing costs by an order of magnitude. Representative models included the IBM 360 series, RCA Spectra 70 series, Burroughs 6500 system, NCR 500, CDC 6000 series, and Digital Equipment PDP-6.

The fourth-generation computer systems introduced in the 1970s feature large-scale integrated circuitry in logic and memory components, standardized communications systems, networking, remote diagnostics, mass storage, data base orientation, and distributed processing. These systems offer a continuing decline in the cost of calculations coupled with increased computing speed. As hardware costs have declined, the design and manufacture of the fourth-generation systems have become increasingly automated. This development has in turn invited entry by low-cost producers selling mainframe systems that can operate with software written for IBM machines. The so-called IBM plug-compatible manufacturers, including Amdahl, Fujitsu, and Hitachi now account for a growing share of U.S. mainframe computer sales.

Since the early 1970s, minicomputers and microcomputers have become powerful enough to perform many tasks that only large mainframe systems could handle before. The low cost and convenience of these small computers have thus created a new demand for low-priced word and data processing equipment. When the mainframe computer manufacturers delayed their response to this new market opportunity, firms such as Digital Equipment, Data General, Hewlett-Packard, Tandem Computer, and Wang Laboratories were able to make major penetrations into the mainframe customer base by offering products with price and performance characteristics that the mainframe manufacturers did not offer. Moreover, as minicomputer and mainframe manufacturers move into the fast-growing office automation and communications markets, mainframe computers will increasingly play a background role in information processing. As a result, mainframe manufacturers are currently faced with major challenges as they reprogram their product and marketing strategies.[8]

As computer prices have fallen and business applications software has proliferated, sales of low-end microcomputers have been transformed from hobbyist orientation into one of the fastest-growing segments of the computer industry. Pioneered by firms new to the computer industry (e.g., Apple Computer, Tandy Corporation's Radio Shack Division, and Commodore International), personal computer sales increased from $77 million in 1976 to an estimated $4 billion in 1982. One forecast predicts that sales of personal computers will reach $20 billion by 1986.[9] Attracted by burgeoning demand, recent entrants include major computer and business equipment firms, semiconductor manufacturers, and a host of small, newly formed companies.

While currently emphasizing the small business user, those personal computer firms that have begun to establish distribution networks are positioning themselves for the future growth of the home information market. As the home information market evolves, it will encourage entry by firms outside the computer industry. Consumer acceptance of these systems will be determined not only by ease of use but also by the ability to perform a wide range of functions such as security, environmental management, home entertainment and education, and access to remote information services. Thus, competitors may come from industries as diverse as consumer electronics, telephone equipment and services, and cable television.

These changes in the relative growth rates of different segments of the computer industry have been marked by the emergence of small, young firms, many of which have introduced such technological innovations as the minicomputer, the microcomputer, and a wide variety of storage and terminal devices. In 1967 the top 50 firms represented 98 percent of the value added of the industry; by 1977 this share stood at 85 percent. Since 1977, the computer industry has witnessed an unprecedented number of startups, spurred by expanding demand, readily available venture capital, steady reduction in component costs, and the development of standard software systems.

BASES FOR COMPETITION

Competition among computer firms has traditionally focused on hardware product features, including quality and reliability, supporting services, and price. Increasingly aggressive price competition has quickened the pace at which technological advances have been incorporated into new product offerings. Price competition is also driving an industrywide move to develop low-cost product and distribution techniques. At the same time, however, emphasis is shifting to software. Fueled by user demands for complete solutions to problems rather than just computing tools, computer firms are turning their attention to more sophisticated systems software. Even minicomputer and microcomputer manufacturers, which formerly emphasized the hardware features of their machines, are now stressing the software support behind their products. Each of these bases for competition--software, production, and distribution--is examined below.

Software

Software programs, or the instructions that guide equipment through its tasks, are the necessary partners in any computer system. Because of the increasing complexity and cost of creating software, many manufacturers have gradually separated the price of software from that of equipment, so-called unbundling.[10] This unbundling has spread from applications programs, written for particular user-oriented tasks such as payroll accounting, to systems software, the programs governing the management of the computer system.

During the 1980s, software is expected to receive greater emphasis both from the user and from computer firms. Users will be seeking ever more sophisticated applications, exemplified by such trends as computer networking and distributed processing. In responding to these market forces, computer firms are directing more of their internal resources into software development. Most firms already have more than half of their development staffs working on software.

The U.S. computer industry currently enjoys a wide lead over the Japanese in software technology.[11] Except in Japan itself, Japanese manufacturers rely on software purchased from U.S. software houses or hire U.S. programmers. Indeed, the Japanese manufacturers' difficulty in developing software is an important factor in explaining why the Japanese have lagged in building computer sales in the United States.

Both the Japanese government and the computer industry, however, have moved to bridge the software gap.[12] Since 1979 MITI has undertaken several important software policy initiatives, including (1) the establishment of a government-funded software technology research center, (2) grants to computer manufacturers to develop new operating systems software, and (3) grants to independent software houses for the creation of new applications software packages, as well as a 40 percent tax deferral on software revenues for the first four years of a program's life. At the same time, Japanese computer manufacturers have increased the magnitude of financial and manpower resources committed to software development. This increased commitment is reflected in the expanding number of software subsidiaries opened by major Japanese manufacturers as well as in the rapid growth of independent software houses in Japan.

The success of the Japanese computer industry in overcoming the software development problem is likely to vary by industry segment. In personal computers, where the competitive opportunities faced by the Japanese manufacturers are ranked highest, software is increasingly being written in standard ways that make it relatively easy to use on different machines. For example, the widespread adoption of the CP/M operating system designed by Digital Research made the CP/M a de facto standard for the first generation of 8-bit microcomputers. By adopting standard operating systems, the Japanese manufacturers of microcomputers will thus be able to avoid an enormous investment in software, since their customers have access to thousands of off-the-shelf programs written for computers made by other manufacturers.[13] While this strategy is likely to facilitate their efforts in the United States, reliance on a standard operating system will not differentiate the personal computers of Japanese manufacturers from those of their competitors.

In large computers, most Japanese manufacturers accept IBM operating systems as a standard and design central processors and peripheral products able to run programs designed for IBM machines. In the United States, for example, Hitachi markets its hardware through National Advanced Systems, a U.S. subsidiary of National Semiconductor that does much of the work in making the Hitachi computers compatible with IBM programs. Similarly, Fujitsu sells hardware to Amdahl, a company that pioneered the IBM-compatible computer concept. To date, however, the program-compatible strategy of the Japanese has enjoyed its greatest success in Japan. In 1979, for example, Fujitsu is reported to have surpassed IBM's Japanese subsidiary in sales volume.[14] Worldwide, program-compatible computers--most of which are made by Japanese firms or include a large proportion of Japanese parts--are estimated to have more than doubled their share of mainframe sales to 19 percent from 1977 to 1980.[15]

The IBM-compatible strategy, however, may not be viable in the long term, particularly if IBM adopts a new computer architecture or lowers its price umbrella. In 1979, for example, IBM introduced its 4300 series computers at a price that competitors found difficult to match--a move generally regarded as a response to potential competition from the Japanese. Another threat to the viability of the Japanese IBM-compatible manufacturers is the possibility that IBM will change its basic operating system. Such a departure has often been discounted since IBM and its customers have more invested in software than hardware. Nevertheless, both IBM and its customers would likely be willing to render obsolete their software investment in the current architecture for a radically new and better computer.

Although IBM-compatible suppliers have shown an impressive ability to adopt the standards established by the system suppliers, system manufacturers are likely to increase the sophistication and complexity of standards in ways that will make achieving this compatibility increasingly difficult. In particular, the trend toward the use of microcode (rather than wired logic) to run a machine's operating system makes it possible to change a machine's nature after it has been installed; this ability provides new degrees of flexibility for the system manufacturer to change and enhance its interconnect standards. For example, IBM's series 3081 computers, introduced in late 1981, contain a new internal structure, called XA for "extended architecture," that allows for post-shipment microcode changes that will be time-consuming, difficult, and expensive for IBM-compatible manufacturers to match.[16]

The Japanese have been least successful in penetrating the fast-growing market for minicomputers, where there is little standardization of software, and proprietry software is developed internally. Like mainframe customers, minicomputer users typically require customized software for many applications, but they want it at a lower price. Accordingly, the profit margins derived from the smaller systems are usually not large enough to justify the investment in software development that the Japanese manufacturers would need to achieve a substantial market share.

Even in Japan, Digital Equipment, Hewlett-Packard, and Data General are the largest suppliers of minicomputers, just as they are in the United States. Although NEC Information Systems, a U.S. subsidiary of Nippon Electric, has marketed its Astra line of minicomputers in the United States since 1979, almost all of the software is produced by independent U.S. companies that buy the equipment from NEC and then write the software. (NEC estimates that 35 percent of the total value of the complete Astra system is produced in the United States.) Reliance on outside software, however, raises the problem of quality control--a crucial issue in a market that increasingly demands high-quality software.[17]

Finally, U.S. researchers currently lead the Japanese in the development of software for the artificial intelligence (AI) systems that are expected to become the basis of the "second computer age."[18] A growing number of U.S. companies, including Hewlett-Packard and Digital Equipment, have established AI research laboratories, while several leading AI scientists have started their own companies. Companies testing AI expert systems include Digital Equipment, IBM, Bell Laboratories, Xerox, and Texas Instruments. Nevertheless, the aim of Japan's recently announced 10-year, $450 million Fifth Generation Computer Project is to develop a prototype of a new family of machines designed especially for AI applications.[19] If successful, Japanese computer manufacturers would establish themselves as leaders in the commercialization of AI and, thus, leapfrog U.S. competitors. While many observers believe the Japanese goals are highly, if not overly, ambitious, the program is likely to give the Japanese computer industry a major competitive boost, even if the project falls short of its objectives.

Production

The computer industry has been transformed in recent years by the declining cost of computing power. As discussed in Chapter 4, the manufacturing costs of semiconductor memory and logic have dropped rapidly over the past decade. Moreover, the costs of some major peripheral equipment items, particularly disk drives, have dropped almost as rapidly.[20] The intensity of competition in the computer industry has, in turn, forced manufacturers to pass through much of the decline in component costs to consumers in terms of increasing performance-price ratios for new models (and selective price cuts for old ones).

While increasing performance-price ratios have created an ever-expanding market, they also mean that a firm must be a low-cost producer in order to be a viable competitor. Mainframe computers that were once low-volume, hand-crafted products with high profit margins have become high-volume, mass-produced products with slim profit margins. Since a manufacturer cannot prevent competitors from copying its technological advances, one of the chief selling points of a computer is the amount of computing power offered at a given price, a factor that is highly dependent on production efficiencies. Indeed, the importance of production efficiencies was dramatically illustrated in 1979 when IBM introduced its 4300 series of mid-sized mainframe computers at a price so low that the resulting competition cut severely into the profits of plug-compatible machines.

In the face of the changing economics of the computer industry, U.S. mainframe manufacturers are investing heavily in automating their factories. Moreover, as the proportion of custom circuitry in computers has risen, manufacturers have also increased their investment in custom logic chips necessary to improve the performance-price ratios of their machines. Shortages of custom chips or poor quality can bring automated, assembly-line production to a halt. Indeed, problems with an industrywide shortage of semiconductor memory and in-house production of proprietary circuits contributed to the widespread production delays experienced by computer manufacturers in 1980.[21]

The shift to low-cost, high-volume computer hardware, particularly in the modules at the heart of the newer IBM computers, also requires a very large computer-based system and equipment design capability. Cost pressures are thus likely to render it increasingly difficult for even a large competitor to continue designing and manufacturing a full range of products. Moreover, the product lines of computer manufacturers increasingly share keyboards, memories, and other major components as manufacturers seek to achieve economies of scale. Computer manufacturers are therefore choosing to buy components of subsystems from other suppliers, to establish specialized, jointly owned subsidiaries (e.g., Computer Peripherals, Inc., jointly owned by Control Data, NCR, and International Computers Ltd.), or to establish cooperative arrangements, particularly on a multinational basis. Even IBM has tended to buy components and peripheral devices from other suppliers more than it has in the past (e.g., printers from Qume and Dataproducts) and, indeed, buys most of the parts for its personal computer.

The transformation to high-volume, assembly-line production has not been an easy one for the computer industry. As they have raced to introduce new equipment with state of the art technology, U.S. manufacturers have experienced problems in the production process not only for mainframe computers but also for data storage devices, minicomputers, and computer terminals.[22] The problem lies in the increasing complexity of custom logic circuits and the resulting increase in the difficulty of correcting design flaws and software problems. The potential cost of such delays is, of course, magnified by the growing number of competitors in the market. The new economics of the industry have thus increased the importance of producing cheaply with a quick design turnaround time.

One possible outcome of the transition to a high-volume, mass-market business is that those manufacturers that cannot make the large investment in plant and equipment may not survive. IBM, for example, invested more than $4 billion in new plant and equipment between 1976 and 1980. At the same time, IBM's aggressive response to the plug-compatible manufacturers--rapid product change and price-cutting--has eroded the price and performance advantages that non-IBM-compatible mainframe firms have traditionally had over IBM.[23] The changing economies of the computer industry have thus begun to cut into the profitability of second-tier mainframe competitors such as Burroughs, Sperry Univac, NCR, and Honeywell. Some observers predict that the number of fully integrated computer manufacturers will be substantially smaller by the end of the decade.

The shift to low-cost, high-volume computer products, however, may assist Japanese computer manufacturers in winning U.S. market share. Just as in the automobile and consumer electronics industries, the traditional manufacturing skills and reliability of Japanese firms give them an advantage at the low-priced end of the product spectrum, which can later be used as a basis for launching a competitive assault on large systems markets where profit margins are higher. The Japanese are expected to be strong competitors in electromechanical peripheral devices such as printers, display terminals, and disk-memory storage devices. Japanese manufacturers have already captured more than half of U.S. sales of low-priced original equipment manufacturer (OEM) printers.[24] Moreover, the Japanese are expected to move aggressively into personal computers where they enjoy low costs for components such as semiconductors and keyboards.[25] Matching the low cost of Japanese manufacturers will be critical for U.S. personal computer manufacturers.

Distribution

Over the past three decades, U.S. mainframe manufacturers have succeeded in building strong marketing and service structures. For this reason, as well as the intricate and deep dependency of users on system-specific software, brand loyalty is probably stronger in large computer systems than in any other segment of the industry. Indeed, no firm in the industry inspires the loyalty that IBM does with its almost legendary service and support.

Given the cost of establishing new marketing service systems, most foreign firms seeking to sell larger computer systems in the United States have used third-party vendors to distribute their products. Of the Japanese computer manufacturers, only NEC has been willing to incur the investment necessary to set up a U.S. subsidiary with direct sales, maintenance, and support staff. In contrast, both Fujitsu and Hitachi have chosen to enter the United States in partnership with a U.S. company.[26]

While the use of partners, third-party system houses, and dealers has helped the Japanese plug-compatible manufacturers to build market share without the expense of setting up a sales network, long-term success will require an investment in building a direct sales organization. However, the cost of such marketing organizations would add to the cost of Japanese computers. Ultimately, this will raise strategic problems for Japanese manufacturers, who use the fact that their products cost less than IBM's as a major marketing tool.

In the low-priced minicomputer and microcomputer product areas, computer hardware costs have dropped so dramatically--due to reductions in the costs of integrated circuits--that it is not possible to follow the traditional industry practice of selling by direct sales contacts and still maintain profit margins. Computer manufacturers have thus intensified their efforts to develop low-cost channels of distribution for low-priced products.

Establishing low-cost distribution channels is perhaps the greatest problem faced by manufacturers of personal computers. Traditionally, computer manufacturers have sold through their own direct sales forces or through third-party system houses. These marketing strategies, however, are too costly when distributing products that sell for less than $10,000. Accordingly, manufacturers of personal computers are experimenting with a variety of selling outlets and techniques, including office equipment dealers, consumer electronics stores, computer specialty stores, department stores, and even mail-order catalogues.[27] Increasingly, however, personal computer retailing is splitting into two major channels: general-merchandise stores and consumer electronics stores, which sell low-cost, popular home equipment, and specialty stores, which offer more sophisticated products and service for business users.[28]

As more manufacturers have introduced personal computers, the industry has become increasingly distribution- and service-limited. Manufacturers relying on independent retailer distribution channels have in some cases been limited by constraints in the availability of dealer shelf space. Even Xerox, which introduced its personal computer in 1981, has reportedly been faced with a shelf-space squeeze. In the home computer segment of the market, however, retail chains--from toy stores and discount houses to television rental operations--are expanding to take up the slack as prices fall.

Although the retail distribution channel for personal computers is closer to that of consumer video and audio equipment--products in which the Japanese are strong--than it is to large computers, personal computers involve considerably more education, training, software, and post-sale support. The best distribution channels for the provision of such support are thus still being explored. One consequence has been that Japanese entry in the United States has moved slowly as the Japanese learn about the market.[29]

Distribution cost pressures have also increased for minicomputers and superminicomputers. In the 1970s minicomputers typically sold through computer systems houses or original equipment manufacturers (OEMs).[30] (OEMs, often small, entrepreneurial concerns, act as middlemen by designing computer software, acquiring peripheral equipment, and selling complete systems to their customers.) Reliance on the OEMs allowed minicomputer manufacturers to avoid the expense associated with the sales and service organizations required to sell directly to customers who demand complete, fully assembled systems. Increasingly, however, minicomputer buyers are bypassing the OEMs to avoid a markup on their computer systems. To do so, minicomputer manufacturers are moving to expand the capability of their sales and service organizations to deal with relatively unsophisticated customers instead of the technologically astute OEMs.

THE CHANGING ECONOMICS OF THE COMPUTER INDUSTRY

The traditional competitive strength of the leading firms in the U.S. computer industry has been based on their ability as low-cost producers with strong customer relationships to develop and sell technologically innovative hardware. Marketing and distribution have been important, facilitating sales with the assurance that customers would receive adequate service. Nevertheless, the market leaders have been those that drove down the price of hardware and added new features.

The nature of competition in the industry, however, is changing as the economics of volume production have turned products such as computer memories and microcomputers essentially into commodities where only a handful of companies can profit over the longer term. In a growing number of markets, state of the art hardware is less important than providing a total system solution specifically tailored for the customer's needs. Today software typically accounts for 50 percent of a computer product's development budget and can go as high as 80 percent.

Computer firms have thus begun to focus on identifying specific customer needs and designing software to fulfill them. In contrast, hardware designs and development have become less important as the significance of a competitor's introducing slightly better hardware has declined. Indeed, some new computer firms have decided to go outside for almost all their components, concentrating their attention on software, sales, and service. Even IBM has increased its purchases of outside components and equipment. Increasingly, success will go to the companies that offer the most useful software and the best customer service and support, not necessarily the most powerful hardware or the lowest price.

NOTES

1. U.S. Department of Commerce, U.S. Industrial Outlook 1982, and "U.S. markets: Data processing and software," Electronics, 13 January 1982.
2. U.S. Department of Commerce, U.S. Industrial Outlook 1982.
3. Ibid.
4. U.S. Department of Commerce, U.S. Industrial Outlook 1981.
5. U.S. Department of Commerce, U.S. Industrial Outlook 1982.
6. U.S. Department of Commerce, U.S. Industrial Outlook 1981.
7. See, for example, "Japan's strategy for the 80's: A worldwide strategy for the computer market," Business Week, 14 December 1981.
8. "Moving away from mainframes: The large computer makers' strategy for survival," Business Week, 15 February 1982.
9. Sales of home computers are still small compared with sales to business, but they are growing rapidly. One estimate is that computers were in only one-half of 1 percent of U.S. households in 1981, but will be in 2 percent by the end of 1982. While one-third of the computers now in homes are the more-expensive models, most new growth is expected to come primarily in the less-expensive models. See, for example, "The home computer arrives," The New York Times, 17 June 1982; "U.S. markets: Data processing and software," Electronics, 13 January 1983; U.S. Department of Commerce, U.S. Industrial Outlook 1982.
10. IBM has led the industry in introducing separate pricing of software. However, while unbundling of hardware and software began over 10 years ago, software prices remain below costs--a factor that has exacerbated the recent earnings problems of mainframe computer manufacturers. See "Computers: Falling behind in mainframe output," Business Week, 20 October 1980.
11. The Japanese lag in software development is attributable to several factors. Language presents a software problem because the Japanese normally write by using a minimum of 2000 Chinese ideograms--or kanji--too many for normal computer programs to accommodate. As a result, programming is done with English words and English versions of Japanese syllables. Japanese productivity in software thus lags the United States.

The Japanese lag in software development is also linked to the slow development of an independent software industry in Japan. Recently, however, the growth of independent software houses has accelerated, spurred both by the unbundling of Japanese hardware and software systems and by the increased rate at which Japanese computer manufacturers have subcontracted software development projects to outside firms.

12. See "Japanese strategy for the 80's: Attempting to overcome the U.S. lead in software," Business Week, 14 December 1981; and "Japan: A big push to rival the U.S. in software," Business Week, 5 October 1981.

13. The Japanese did not appreciate the importance of applications programs when they designed their first-generation desk-top computers with 8-bit microprocessors. Most of the new generation of Japanese 16-bit microcomputers will be able to use the same software as IBM's personal computer. See, for example, "Japan: A lame debut for personal computer exports," Business Week, 14 June 1982.

14. "The 'Japanization' of an IBM subsidiary," Business Week, 6 April 1981. It should be noted that IBM moved early to establish a position of strength in the Japanese mainframe industry before two key competitors, Fujitsu and Hitachi, could gain dominance. Still holding almost 25 percent of the market, IBM is denying its Japanese competitors both production experience and cash to develop the distribution and software capabilities essential to success in the United States.

15. "Moving away from mainframes: The large computer makers' strategy for survival," Business Week, 15 February 1982.

16. The demand for information about technological developments in IBM's computer architecture was underscored recently by charges that Hitachi, one of Japan's leading computer companies, conspired to buy secrets stolen from IBM. See, for example, "Competitors have sought IBM's secrets for years via courts, research, hiring," The Wall Street Journal, 25 June 1982; "IBM watchers process data on the big firm to divine its program," The Wall Street Journal, 23 July 1982; "Computers: IBM's mimics struggle to keep pace," Business Week, 20 September 1982.

17. For example, reliance on outside applications software providers has caused well-publicized problems for Prime Computer. See, "Computers: A big challenge at Prime Computer," Business Week, 21 December 1981.

18. See, for example, "Artificial intelligence: The second computer age begins," Business Week, 8 March 1982.

19. The Japanese have special reasons for pursuing AI. First, the written Japanese language poses severe problems for the Japanese in attempting to close the software gap with the United States. The idea of an AI-based fifth-generation computer thus

has special appeal to the Japanese. Second, in the mid-1970s, MITI directed Fujitsu, Hitachi, and eventually Mitsubishi into making computers to run on IBM software. This symbiotic strategy helped the Japanese industry compete with IBM's System 370, but increased its vulnerability whenever IBM launched new products or cut prices. The fifth-generation project's unconventional goals thus offer a means of reducing dependence on IBM. Finally, due in part to over $125 million in subsidies from MITI for its VLSI program, Japanese electronics firms are now at least as skilled as their U.S. counterparts in manufacturing densely packed integrated circuits that are essential to fifth-generation computers. See, for example, "A fifth generation: Computers that think," Business Week, 14 December 1981; and "Japan: Here comes Computer Inc.," Fortune, 4 October 1982.

20. Magnetic disk storage units, which act as slower, auxiliary storage to high-speed semiconductor main memory, have substantially increased in performance and decreased in price for several decades, with order of magnitude improvements still anticipated. For example, since the mid-1950s when disk storage units began to appear, storage capacity, as measured by the number of megabytes per spindle, has increased by an average of about 25 percent per year. In contrast, price, as measured by monthly base charges per megabyte, has declined by almost 20 percent per year.

21. Also contributing to production delays in 1980 were producer miscalculations with respect to the magnitude of customer demand.

22. These production glitches have delayed new product delivery schedules by as much as a year for firms ranging from IBM, NCR, Honeywell, Burroughs, Digital Equipment, and Magnuson, to Apple Computer. See, for example. "Computers: Falling behind in mainframe output," Business Week, 20 October 1980; "Computers: Snafus that delay new products," Business Week, 1 June 1981; "Computer design may save Magnuson," Business Week, 15 February 1982.

23. See, for example, "Moving away from mainframes: The large computer makers' strategy for survival," Business Week, 15 February 1982; "IBM's aggressive pricing," The New York Times, 9 August 1982; "No. 1's awesome strategy," Business Week, 8 June 1981.

24. See, for example, "Peripherals: Japan's swift success in printers," Business Week, 31 August 1981.

25. For example, low-power consumption C-MOS semiconductors will be key components in the design of personal computers that are portable, compact, and battery-driven. Although C-MOS semiconductor technology was originally advanced by U.S. semiconductor manufacturers, several Japanese manufacturers

have reportedly made major investments in C-MOS assembly lines and are said to be well down the C-MOS learning curve. "Information processing: A worldwide strategy for the computer market," Business Week, 14 December 1981; "The big fights for computer sales," The New York Times, 1 August 1982; and "New integrated circuits may spur semiconductor industry," The Wall Street Journal, 18 February 1982.

26. Fujitsu and Hitachi have followed a similar mainframe sales strategy in Europe, where Fujitsu has established technology and marketing agreements with ICL in Great Britain and Siemens in West Germany, and Hitachi with Olivetti in Italy and BASF in West Germany.

27. IBM, which introduced its personal computer in mid-1981, is investing heavily to test alternative distribution channels. Its distribution strategy calls for a three-pronged marketing approach: use of its direct sales force to sell personal computers to corporations for use by managers; distribution through independent retailers, who are supported by point-of-sale materials, warranty reimbursements, and a hotline to answer questions; and establishment of a company-owned chain of computer stores to attract small business owners. "IBM joins the race in personal computers," Business Week, 24 August 1981.

28. For example, Tandy Corporation's Radio Shack Division is currently opening a new chain devoted exclusively to computer sales in addition to its chain of consumer electronics supermarkets.

29. To date, Japanese manufacturers of personal computers have concentrated on their home market, where they had achieved an estimated 75 percent share in 1980. Their success reportedly stems from their pricing strategies, the inability of their U.S. competitors to produce enough units to meet the rapid expansion in Japanese demand, and the lack of U.S. products capable of storing and retrieving the approximately 2000 characters of the written Japanese language. U.S. Department of Commerce, U.S. Industrial Outlook 1982; and "The big fight for computer sales," The New York Times, 1 August 1982.

30. An exception to the general reliance on OEMs in the 1970s was Prime Computer, which conceived a strategy of selling directly to computer users.

6
The Telecommunications Equipment Industry

In the 1970s, the telecommunications equipment industry entered into a period of accelerating change, mainly as a result of developments in the industry's technology. New products and technologies have led to rapid growth in private equipment markets, and new export markets for telecommunications equipment have emerged. Moreover, as markets have opened up, new firms have been encouraged to enter the equipment industry, thus accelerating the innovation process. Even in those parts of the equipment market dominated by telecommunications providers, new technologies have led the service providers to expand their range of suppliers.

This chapter examines the influence of these developments on the structure of the U.S. telecommunications equipment industry and its international competitive position, with special emphasis on switching equipment. The organization of the chapter is as follows. The first two sections provide an overview of the U.S. industry and its international position, including problems faced by U.S. firms in world markets. In contrast to the semiconductor and computer industries, U.S. telecommunications equipment manufacturers have primarily been oriented to domestic sales. This orientation is attributable in part to barriers created by the structure of national telecommunications service systems and related political and national security considerations. The low relative U.S. export share is also due in part to the fact that, until recently, Western Electric, the largest U.S. manufacturer, focused almost exclusively on domestic sales to the Bell System operating companies.

The third section discusses major developments in transmission technology. The introduction of microwave, satellite, and optical fiber transmission has had important effects on industry structure. In addition, the shift to digital transmission in telephone plants laid the groundwork for digital switching, which has emerged as an area of intense international technological rivalry and competition.

The fourth section analyzes the bases for competition in central office switches and private branch exchanges (PBXs). The section's conclusions are similar in three aspects to those found in the previous chapter on the computer industry. First, rapid change in digital integrated circuits has promoted change in the structure of the switching equipment industry. Second, software is an increasingly important determinant of switching equipment competitiveness. The success of U.S. manufacturers is directly linked to the strength of the U.S. industry in software development. Indeed, the most successful foreign switching equipment firms competing in the United States have established U.S. design and manufacturing operations. A third similarity with the computer industry is the critical role that distribution and service networks play in competitiveness. Foreign firms seeking to enter the United States thus face the requirement for an investment in distribution and service capabilities, the magnitude of which may serve as a barrier to entry.

INDUSTRY SIZE AND INTERNATIONAL POSITION

The equipment used in telecommunications networks is traditionally classified into three functional groups. Terminal equipment, normally located on the customer's premises, is used to originate and receive signals. Transmission equipment carries the signals between terminal stations and central offices and between switching centers. Switching equipment, located in central offices and on customer premises, links the terminals or switching nodes in the network.

Because the demand for telecommunications equipment is a derived demand, dependent on the demand for telecommunications equipment services, the organization of the telecommunications services industry has a major impact on the structure of the equipment industry.[1] In virtually every developed country, the provision of telecommunications services has historically been organized as a monopoly. Equipment sales within countries have also been highly concentrated with the share of the four largest firms in industry sales typically being above 70 percent. In contrast to other electronics-based industries, small, specialized firms have played only a minor role in the provision of telecommunications equipment. Through their procurement practices, service providers have traditionally acted to perpetuate this industrial structure, protecting equipment suppliers from both domestic and foreign competition.

The United States is the world's largest consumer of telecommunications equipment and services. In 1972 the value of U.S. shipments of telephone and telegraph equipment was $4.5 billion; by 1981 the rapidly evolving information age had pushed the value of industry shipments up to $12.2 billion.[2] The industry's growth in the 1980s will be sustained by the continuing transition from the present partially analog, partially digital telecommunication network to all digital. The pace of this change will be set by the economics of furnishing traditional telephone service, which will remain dominant over the next decade in terms of common carrier investment and revenues.

U.S. exports of telephone and telegraph equipment were estimated at $653 million in 1981, up 17.2 percent from $557 million in 1980.[3] Similarly, U.S. imports were approximately $495 million in 1981, up 17.6 percent from $421 million in 1980. The U.S. telecommunications equipment trade surplus in 1981 was thus $158 million, up 16 percent from the 1980 surplus of $136 million. Table 6-1 compares the telephone and telegraph equipment and parts exports of the 10 principal competitor nations.

Historically, telecommunications equipment manufacturers in most countries have primarily been oriented to domestic consumption, with exports accounting for a small part of total output.[4] In the 1970s, however, exports of telecommunications equipment manufacturers rose dramatically. Many of the major countries of destination for the European exporters in 1980 were developing nations, some of which have launched multimillion dollar projects to modernize and expand their communications infrastructures.[5] These include Saudi Arabia, South Korea, Taiwan, Mexico, and Argentina. By contrast, about 70 percent of Canada's exports and 32 percent of Japan's exports went to the United States. The principal countries of destination for U.S. exports were Canada, Taiwan, and the United Kingdom.

Western Electric, the largest supplier of telephone and telegraph equipment in the world, has traditionally neither exported nor imported its products. Until recently, Western Electric directed its entire productive capacity toward meeting the requirements of the Bell System operating companies and the U.S. government.[6] Thus, while Bell System innovations have offered continued improvement in the cost, quality, and availability of service in the United States, the overall impact of U.S. telecommunications technology in world markets has been low relative to the high level of domestic telecommunications development. In contrast, telecommunications equipment manufacturers in West Germany, France, Sweden, the United Kingdom, and Japan have been major exporters.[7]

Although Western Electric is likely to mount an aggressive marketing drive overseas following the implementation of the

TABLE 6-1 Telephone and Telegraph Equipment and Parts Exports of Principal Nations

Principal Nation	1980 Exports (millions $)	Share of Total Exports (%)	Exports Growth Rate 1977–1980 (%)
West Germany	878	19	7.9
Sweden	809	18	13.9
United States	557	12	29.4
Japan	538	12	15.1
The Netherlands	466	10	17.1
Belgium/Luxembourg	390	9	7.4
France	327	7	14.6
Canada	252	5	45.4
United Kingdom	223	5	–9.2
Italy	144	3	4.0
Total	4,584	100	12.9

SOURCE: Official government statistical publications of each country. As presented in the U.S. Department of Commerce, *U.S. Industrial Outlook 1982* Washington, D.C.: U.S. Government Printing Office.

AT&T Consent Decree of 1982, foreign telecommunications manufacturers will remain strong competitors. The nature of national telephone systems is such that they are frequently constructed in building-block fashion, that is, they often start with the basic telephone and telegraph equipment and build up to the complex, sophisticated, total telecommunications network. Once the basic equipment has been purchased, the follow-on equipment must be compatible (i.e., manufactured to the same specifications). Therefore, it is frequently furnished by the original manufacturer.

Another problem faced by U.S. manufacturers selling abroad arises from the differences between North American technical standards based on Bell System practices and the International Telephone and Telegraph Consultative Committee (CCITT) recommendations used in most countries, with the major exception of Japan.[8] The significance of differences in technical standards, however, will diminish as U.S., Japanese, and Canadian manufacturers play an increasingly active international role and as CCITT-based manufacturers seek to sell in the United States. Some companies, for example, are attempting to design switching systems in which the bulk of the equipment is compatible with either CCITT or Bell standards.

Finally, protective policies and practices pose significant barriers to importation of U.S. telecommunications equipment into many foreign countries. In most countries, with the major exception of Canada, the telephone and telegraph system is government-owned and operated. Foreign telephone systems, particularly in Europe, the United Kingdom, and Japan, traditionally have channeled their purchases to favor their own national economies, i.e., to locally owned companies or to foreign-owned companies with local manufacturing facilities.[9] Even the developing countries have established "buy local" policies as they have generated enough demand to support local manufacture.[10]

Although protective policies have largely precluded the importation of U.S. telecommunications equipment where there is competition from local manufacturers, U.S. firms have been able to export domestically produced equipment in specialized high-technology areas. U.S. telecommunications equipment firms have also enjoyed marketing success in foreign countries where they have established manufacturing facilities. For example, foreign plants are the bases of the foreign sales volume of ITT and GTE.[11]

Mutual reciprocity between trading partners may increase as foreign telephone authorities in many countries open sales of customer equipment.[12] For example, the British and Canadian governments recently took steps to open customer equipment sales to foreign manufacturers. Sales of interconnect equipment may eventually provide an avenue for direct sales to the postal, telegraph, and telephone authorities overseas. Another potentially significant development is the three-year bilateral agreement signed by Japan and the United States in 1980 to permit U.S. suppliers to compete with Japanese manufacturers and other foreign countries in the supply of equipment to Nippon Telegraph and Telephone.[13]

INDUSTRY STRUCTURE

The structure of the U.S. telecommunications equipment industry closely parallels the structure of the telephone service industry. Traditionally, both the Bell System and GTE have relied heavily, though not exclusively, on their captive suppliers. AT&T, which serves over 80 percent of the country's telephones, accounts for a proportionately large share of equipment sales through its manufacturing subsidiary, Western Electric.[14] GTE, with approximately 10 percent of the nation's phones, accounts for a similar proportion of equipment sales through its manufacturing arms.

The remaining portion of the U.S. industry is composed of a number of independent manufacturers. Suppliers offering broad lines of telecommunications equipment include ITT, Northern Telecom, and NEC America. ITT is the second largest producer of telecommunications equipment in the world. However, more than two-thirds of its sales and manufacturing operations are in western Europe. Northern Telecom, the manufacturing subsidiary of Bell Canada, has substantially expanded its U.S. operations in recent years, particularly its marketing of digital central office switches. NEC recently established U.S. production facilities for central office switches and PBXs, though it imports other products ranging from key telephone systems to microwave equipment. Other major suppliers of telecommunications equipment in narrower product areas include Collins Radio, Farinon Electric, Lynch, and TRW Vidar in transmission equipment; Plessey in switching equipment; and ATI/Fujitsu, Mitel, ROLM, and TIE in PBXs and key telephone systems.[15]

The non-affiliated equipment manufacturers have historically sold largely to independent telephone companies and private buyers. The latter group of customers has been a source of growing overall market share for independent manufacturers since the 1968 Carterfone decision permitted interconnect sales. In 1969 the Bell System and the independent telephone companies together accounted for nearly all of the telephone equipment purchased in the United States; by 1981 sales of telephone equipment to private buyers accounted for over 10 percent of total industry sales.[16] The independent manufacturers have also increased sales to Bell and GTE operating companies, and their share may increase further in the future with the divestiture of the Bell operating companies from AT&T.[17] However, the independents will face increased competition for their traditional customers from Western Electric, which, following the 1982 Consent Decree, has new incentives to sell to non-Bell companies as well as to enter markets other than telecommunications equipment.

DEVELOPMENTS IN TRANSMISSION TECHNOLOGY[18]

Several important changes in telecommunications technology have affected industry structure and competitiveness. One of the most recent--the shift to digital switching--is the subject of a case study in the following section. However, the groundwork for the shift to digital switching was laid by the shift to digital transmission in the telephone plant that began in the early 1960s when digital trunk carrier systems were introduced. At first, these systems were installed in metropolitan-area telephone networks, where they relieved trunking shortages without the high cost of installing additional cable pairs in cities. Later it was found that the economics of digital transmission could also be realized in medium- and long-haul trunking. Today, a growing proportion of the intercity and local interoffice transmission in major national telephone networks is performed by digital carrier systems.[19]

The potential for explosive growth in digital data transmission has also contributed to the incentive to develop digital switching systems. In addition, data transmission has inspired considerable jockeying for competitive position among computer and telecommunications equipment manufacturers in areas ranging from local area networks to private branch exchanges. In the 1980s, however, only a minority of subscribers connected to the national telecommunications network will need digital data transmission. Thus, over the near term, digital data service will continue to be provided in many cases by the special treatment of existing local loops, e.g., at the subscriber premises or local exchange.

Although copper cables remain the center of the transmission network, technological advance has expanded the range of transmission media. In addition to the development of the digital trunk carrier, other notable developments in transmission technology include terrestrial microwave and satellite communications, both of which are suited for either analog or digital transmission. These two technologies have promoted entry into long-distance telecommunications service and increased the derived demand for independent manufacturers' equipment. A third development, optical fiber transmission, is notable for its return to a land-based medium and its suitability only for digital transmission. Furthermore, optical fiber technology is expected to produce a new generation of telecommunications switches, based on light rather than electrical impulses.

In the 1940s, microwave developed as an alternative to coaxial cable for long-haul transmission. Microwave transmission has had a major impact on industrial structure. Previously a small number of copper cable producers dominated the point-to-point transmission equipment industry; by 1978, however, microwave radio accounted for over 75 percent of annual expenditure on long-haul transmission equipment by U.S. common carriers.

The development of satellite communications overcame some of the limitations of terrestrial microwave systems, creating a major new competitor to cable and microwave systems in long-haul and intercontinental transmission. Since the 1965 Intelsat-I satellite, space communications costs have steadily declined while the range and quality of space communications services have grown. Although Bell Laboratories was involved in the initial work on satellite communications, satellite technology was essentially developed by U.S. aerospace manufacturers under contract from the U.S. National Aeronautics and Space Administration (NASA) and the U.S. Department of Defense. Indeed satellite equipment manufacture is still primarily oriented to military purposes, since military satellites account for 70 percent of all satellite launchings. As a result, the satellite equipment industry is dominated by U.S. aerospace manufacturers with a long history of defense work: Hughes Aircraft, Ford Aerospace, TRW, RCA, and General Electric.

The development of optical fiber transmission during the past decade also promises to affect the structure and performance of the telecommunications industry. The major initial areas of application over the next decade are likely to be interoffice trunks (i.e., the high-density cables linking urban exchanges), long-haul transmission, and local data transmission links to high-volume users. However, optical fibers may also serve as the basis for distributing a broad range of communications signals in addition to conventional telephony, though the economics of broadband distribution systems remain uncertain.

The basic material in optical fibers is silica, which is cheaper and more widely available than the copper used for conventional cables. Moreover, optical fibers are suited for digital pulse code modulation (PCM) transmission. Optical fibers also have a number of major advantages over other terrestrial transmission media: more than three times the capacity of a coaxial cable of given dimensions; greater resistance to corrosion than metal wire, but less attenuation of communications signals so that fewer regenerators are required on long routes; immunity to electric interference, which makes fiber more reliable than metal wire in areas where high current may be passing; and greater security because of the difficulty of illegally tapping the traffic flow.

The production of optical fiber equipment basically involves two product groups, the optical fibers themselves and the optoelectronic components used to transmit and receive signals along these fibers. An early lead in optical fibers was taken by Western Electric and Corning Glass, which hold many of the basic patents on the fiber optic manufacturing process. Recently, however, a number of major telecommunications firms--including AT&T, ITT, NEC, Northern Telecom, and Philips--have developed processes that are reported to be substantially different from those patented by Corning Glass.

Two types of materials are used to transmit signals along optical fibers: light-emitting diodes (LEDs) and lasers. LEDs--an area in which the Japanese hold a technology lead--are primarily produced by electronic component manufacturers, including the component divisions of telecommunications equipment manufacturers. Lasers, on the other hand, are produced by a number of small, highly specialized firms and by space-defense manufacturers. Consequently, it will not be easy for telecommunications equipment manufacturers to directly enter laser production, though some are doing so through acquisitions of specialized firms.

While much important early work in fiber optics was done in the United States, NTT and its Japanese suppliers have played an active role in development work on glass fibers, LEDs, and receivers. Moreover, NTT's closed procurement policies have sheltered its suppliers by excluding foreign firms. At the same time, Japanese firms have been able to build volume by pursuing highly aggressive price strategies for the provision of complete telephone systems in developing country markets outside North America and Europe.[20] The Japanese are thus building an extremely strong position in a product line that is expected to grow rapidly in the next decades.

BASES FOR COMPETITION:
A CASE STUDY OF SWITCHING EQUIPMENT

Switches are in several ways the heart of modern telecommunications systems. The "star" architecture that dominates telecommunications design places a switch at the node where multiple transmission lines intersect. The capacity of these switches ranges from under 100 lines in the smallest PBXs, used for business applications, to more than 100,000 lines in large telephone company central offices.[21] In addition to their physical location in the system, today's switches control most of the functions of modern telecommunications systems. These functions range from traditional ones such as providing dial tone, ringing, and busy signals, to advanced features such as call forwarding and least-cost toll routing.

Since the mid-1960s, electronic switching has found wide acceptance. The earliest electronic switching systems used wired logic--special purpose circuitry, with modules of memory and logic, tailored specifically to the application of the telephone system. Wired logic has now been replaced in new product design by stored program control (SPC) in which computer-like processors perform necessary switching functions. With SPC, changes can be made in customers assignment, class of service, and options by changing hardware connections. The first SPC switching system was introduced by Western Electric in 1965; since then, more than 40 SPC switches have been developed by telecommunications equipment manufacturers around the world.[22] Until the late 1970s, however, most SPC central office switches still utilized electromechanical elements in the switching matrix to achieve connections between telephone circuits.

Several factors combined during the past decade to push telecommunications switching into digital technology. One factor was the rapid decline in cost and increases in performance of digital integrated circuits. Closely related to this was the increasing base of knowledge in computer software techniques. A third important consideration was the widespread use of digital transmission systems, discussed earlier, that provided a more economical interface with digital than analog switching. Although analog switching remains common due to the large installed base of older equipment, most manufacturers' latest product offerings incorporate digital techniques.

Equipment manufacturers' R&D costs have escalated with the transition to digital technology. During the late 1970s, for example, ITT spent some $300 to $600 million on its yet to be completed 1240 switching system.[23] This investment stands several orders of magnitude above the $30 to $40 million that ITT spent in the early 1960s on its Pentaconta switching system--a system whose commercial life was nearly 20 years. Similar investments in the development of fully digital switching systems have been made by telecommunications equipment manufacturers around the world, some of which have spent more than $600 million.[24]

In the face of rising R&D requirements, telecommunications equipment manufacturers have sought to increase their international sales.[25] Moreover, as growth in demand for switching equipment has slowed in major producing country markets, firms that have traditionally oriented their operators toward domestic consumption have intensified their efforts to expand international export market share. Notable examples include Canada's Northern Telecom and Japan's Nippon Electric, Oki, Hitachi, and Fujitsu. Until recently, this strengthened export orientation was also encouraged by strong demand for telecommunications equipment in the developing countries, particularly in the Organization of Petroleum Exporting Countries (OPEC) area.[26] Exports to developing countries, however, are not expected to expand as rapidly in the 1980s as they did in the 1970s.[27]

Competition between switching equipment manufacturers will thus focus increasingly on OECD countries. Further, increased competition will be encouraged by the technological characteristics of electronic switching systems. The flexibility of SPC switching has made it easier to adapt systems developed for one network to the technical characteristics of other networks.[28] However, switching technology will continue to require an understanding of large-scale network behavior that is highly country-specific, and thus a major barrier to entry into national markets will continue to exist.

Competition in the United States will also undoubtedly be intensified by the implementation of the 1982 AT&T Consent Decree. Domestic manufacture, however, will remain far more competitive than importation for three major reasons. First, the design and development of new switching systems is an ongoing process, involving continuing interaction between designers and network managers. Telephone companies cannot limit themselves to screening new equipment once it has been designed; rather they must identify new equipment needs and assist manufacturers in meeting these needs. Moreover, telephone companies require sources of equipment supply that will be available over the long term to supply compatible equipment for subsequent expansion and modernization.

A second factor favoring U.S. manufacture is the rapid pace of technological change in the United States. By the end of 1979, there were approximately 300 million telephone main stations in the world. Of these, 100 million were in the United States. Because of its magnitude, the U.S. telephone system dominates U.S. telecommunications and strongly influences developments worldwide. Consequently, competitors will find it highly advantageous to locate in the United States in order to anticipate technological developments and to design, produce, and successfully market new products that will meet customer needs.

Finally, the economics of the design and production process favor a U.S. manufacturing location over importation.[29] VLSI logic circuits, for which the U.S. semiconductor industry is the world leader, account for an increasing share of the total direct cost of switching equipment. As the number of switching functions integrated into one semiconductor package continues to rise, the component parts of switching equipment will increasingly resemble the finished product. Telecommunications equipment manufacturers seeking to stay close to the forefront of semiconductor technology have thus placed increasing emphasis on integrated circuit design and production, either by acquiring semiconductor manufacturers or expanding their own captive operations.[30]

For foreign firms other than the Canadians and the Japanese, the barriers to importation into the United States are heightened by the differences between U.S. and CCITT technical standards. The importance of local requirements with their strong software component is perhaps the strongest argument for a U.S. manufacturing location. It has been estimated that conversion of switches offered by established European manufacturers would require two to three years of effort. Indeed, in 1979, CIT-Alcatel, the pioneer in European digital switching, initiated a switch development effort in the United States that had not been completed by the end of 1982.

The importance of U.S. manufacturing capacity is demonstrated by the pattern exhibited by foreign firms marketing central office switches in the United States. The largest foreign participants, Northern Telecom, NEC, and Plessey, have established or acquired U.S. manufacturing facilities. Two other entrants, Fujitsu and CIT-Alcatel, have also announced that they will rely on U.S. manufacturing operations. Even ITT, which is a major supplier of switches in western Europe, designs and manufactures switches for U.S. sales domestically.

The design and manufacture of electronic PBX systems also require expertise in hardware, solid-state switching, telephony, and software. As in the case of central office switches, production economics favor U.S. manufacture of PBX equipment over importation, particularly for large systems. The need to anticipate technological development in the U.S industry, as well as to tailor product software to customer needs, is highly important in competing for market share.[31] Foreign firms that have established U.S. manufacturing operations include Northern Telecom, Mitel, NEC, Fujitsu (American Telecom Inc.), Oki Electronics, Hitachi, and Siemens.

The Carterfone decision in 1968 and subsequent FCC rulings permitted sales of terminal equipment directly to customers-- referred to as interconnection--in addition to the existing practice of leasing equipment from the telephone operating companies.[32] Since the Carterfone decision, there has been sustained entry into the industry. In 1969 there were only four PBX manufacturers in the United States; by 1980 there were over 30.[33]

The first wave of entrants were established foreign manufacturers represented by U.S. distributors with responsibility for installation and maintenance. These firms, which included L. M. Ericsson, Hitachi, and NEC, were quick to exploit the relatively sluggish marketing program of the telephone companies as well as the perceived nonmodern nature of U.S. manufacturers' step-by-step PBX equipment. By 1974 foreign suppliers had garnered as much as 16 percent of the interconnect sales.[34]

The invasion of foreign manufacturers slowed in the mid-1970s when the introduction of U.S.-designed electronic systems made the older crossbar technology systems of Japanese manufacturers obsolete. The electronic PBXs of North American manufacturers such as ROLM, Western Electric, and Northern Telecom offered both extensive flexibility and appealing special features and, therefore, were far easier to sell than the more limited crossbar systems. Sales of the less expensive crossbars, however, have continued in less sophisticated applications.

Foreign firms marketing PBX systems in the United States have also found it necessary to invest in building distribution organizations. As in the computer industry, a strong distribution network for marketing and service is essential if a PBX manufacturer is to be a viable competitor. Since telephone switching equipment problems must be repaired immediately, sales and service distributors must be able to guarantee the availability of parts and trained repairmen. Interconnect suppliers must demonstrate that they can supply round-the-clock service and maintenance to match the telephone companies. Such service capabilities require that individual distributors be both well managed and well capitalized.

Almost by definition, the U.S. distributors, which intially marketed the products of foreign manufacturers, were small companies that could only offer regional and, sometimes, unsatisfactory service. When matched against large firms, such as AT&T, ITT, General Dynamics, and GTE, small independent distributors found themselves at a competitive disadvantage. This disadvantage appears to be particularly strong in the competition for large system accounts since national companies show a preference for dealing with a single equipment representative rather than a number of small firms scattered across the United States. Since the mid-1970s, foreign manufacturers have thus attempted not only to catch up technologically but also to improve their U.S. distribution networks.

The changeover to digital technology since the mid-1970s has increased entry into telecommunications switching, resulting in intensified product competition. The competition accompanying the digital changeover has also accelerated the innovation process. For example, in the 1960s, it took six to eight years to develop a new PBX system; by the 1980s the development period was three to five years. At the same time, product lifetimes have declined as new generations of equipment enter the marketplace each year; current trends suggest replacement rates of between 10 percent and the 20 percent level standard in the computer industry.[35]

NOTES

1. For a discussion of the impact of the organization of telecommunications services on the structure of the telecommunications equipment industry in the OECD countries, see Organization for Economic Cooperation and Development, Telecommunications Industry Equipment Study, October 1981, pp. 23-43.

2. Industry shipments increased by only 7.3 percent in 1981 to $12.2 billion, a slowdown that is attributable in part to the impact of high interest rates on plans of telephone companies to convert from electromechanical to electronic switching. U.S. Department of Commerce, U.S. Industrial Outlook 1982.

3. Ibid.

4. The major exceptions are Sweden's L. M. Ericsson and the Netherlands' N. V. Philips, both of which have a limited domestic market.

5. As discussed in the OECD's recent telecommunications study, competition for developing country sales has been greater than in the OECD countries for two reasons. "First, domestic production in developing countries is much more limited, so that preferences for domestic producers have been less of an impediment to trade. Second, existing networks in the developing countries are either very small or almost totally legacies of colonial rule, so that network compatibility problems have arisen to a lesser degree." OECD, op. cit., p. 47.

6. In 1978 Western Electric established a new subsidiary (Western Electric International, reorganized in 1980 as AT&T International, an unregulated subsidiary of AT&T) to sell telecommunications equipment and related services in the rest of the world. Major sales have been made to Saudi Arabia (for construction of a microwave network), South Korea (for local and toll switching systems), and Taiwan (for toll switching systems).

7. Domestic sales available to non-U.S. firms are smaller by several orders of magnitude than those available to firms operating in the United States. Acquiring or defending competitive positions in the smaller OECD countries and in third world countries is therefore a more important objective of their corporate strategy. These firms have rarely sought to compete in other major countries on a broad-line basis since aggressive behavior would expose them to retaliation in their home market. Rather, if they do compete in these countries, it has been on the basis of highly specialized products. OECD, op. cit., p. 34.

8. CCITT standard-setting usually lags behind Bell standards. In some cases the CCITT standards are improvements on earlier Bell standards. In others they may be necessitated by different foreign infrastructures such as metric spacing of repeaters. In still other cases, the foreign differences may reflect economic (e.g., protecting domestic equipment producers) or political (e.g., non-American), rather than technical, considerations. In addition to CCITT standards, foreign countries also have their own national technical standards, which are not uniform.

9. PTTs in Europe and Japan not only buy their equipment from these selected national manufacturers but also set the standards for the equipment and provide financial support for research and development programs.

For example, Japan's Nippon Telegraph and Telephone Public Corporation (NTT) maintains a very close relationship with its Japanese suppliers through joint R&D programs, direct investment in manufacturing companies, and interlocking of senior executives as a result of many ex-NTT staff entering into industry executive positions following retirement from NTT. Moreover, with its historically closed procurement policies, NTT has helped to stabilize prices and production for Japanese manufacturers. NTT also advances its Japanese suppliers part of the equipment purchase price, in effect providing interest-free loans. The result of this government support has reportedly allowed Japanese manufacturers to be highly aggressive in export pricing. See "Japan: The coming assault on communications markets." Business Week, 14 December 1981.

10. For example, in Brazil, Mexico, and Argentina traditional telecommunications equipment suppliers--Ericsson, ITT, and Siemens--are currently being forced to give up control of their in-country manufacturing facilities in order to continue participating in these countries.

11. Due to local manufacturing operations, ITT and GTE have historically enjoyed strong positions in Europe. For example, in 1925 ITT acquired the International Western Electric Company, which held a strong position in European telephony and had developed extremely close ties with service providers. ITT has capitalized on these ties, while attempting to create similar ties in third world countries. Although they have in some cases shared a common core technology, ITT subsidiaries have therefore been relatively autonomous, often acting as domestic rather than international firms. OECD, op. cit., p. 35.

12. For a discussion of moves toward liberalization of trade in telecommunications equipment, see OECD, op. cit.

13. Although NTT has taken steps to open procurement to foreign firms since January 1981, U.S. telecommunications equipment manufacturers have reportedly been highly tentative in approaching NTT. See, for example, "Why the U.S. still has not cracked NTT," Business Week, January 1982. Through mid-1982 the only significant breakthroughs were by Motorola, which has been placed on NTT's approved supplier list for mobile telephone equipment, and ROLM, which has been placed on NTT's approved supplier list for PBXs. "Motorola hurdles a Japanese barrier," Business Week, 7 June 1982, and ROLM Press Release, 29 March 1982.

14. Western Electric's exact share is subject to debate, depending on what types of equipment are included and what data sources are used. Western's share, however, has declined somewhat since the mid-1970s due to factors such as increased competition for sales of equipment to private businesses and the more rapid growth of equipment demand by independent telephone companies (not served by Western) than by Bell operating companies.

15. Many of the independent suppliers are subsidiaries of large corporations, some of which are foreign. Collins is a division of Rockwell International; Farinon, a subsidiary of Harris Corporation; and Vidar, formerly a subsidiary of United Telephone, is now

part of TRW. Lynch is partially owned by the large French supplier, CIT-Alcatel. Stromberg-Carlson, until recently a broad-line
subsidiary of General Dynamics, had its central office and private
branch operations sold to United Technologies, which in turn sold
the central office switching business to Plessey of the United
Kingdom. American Telecommunications, Inc. and Fujitsu have a
U.S.-based PBX joint venture. Mitel, of Canadian origin, manufactures a large proportion of its equipment in the United States.
TIE, a U.S. company, obtains some of its key telephone equipment
from an affiliate in Taiwan. In addition to these companies, several other large foreign companies have gained small U.S. market
shares in microwave equipment and PBXs through imports, and
some, such as Siemens, have U.S. manufacturing and R&D
operations.

16. U.S. Department of Commerce, U.S. Industrial Outlook
1982.

17. Although some observers have raised the issue of the
coordination and overall quality of the nation's transmission
network following AT&T's divestiture of the local operating
companies, this problem does not stem from the identity of the
equipment manufacturers. Rather it arises from the incentives of
equipment purchasers and the fact that the technical quality of
the local telephone network is an excludable public good. "Network degradation" can occur if either (1) local access pricing is
not sufficiently responsive to permit the local operating companies to recoup the benefits of enhanced local network quality,
or (2) antitrust or state regulatory restrictions impair the ability
of the operating companies to jointly determine interconnection/
quality standards. Although the post-divestiture industry structure permits a central staff mechanism for research funding and
coordination, this does not guarantee that problems will not
occur, particularly with respect to local pricing.

18. The description of developments in transmission technology in this section is based largely on OECD, op. cit., pp. 49-58.

19. In subscriber loops the conversion to digital transmission
will proceed rather slowly. There is less technical or economic
incentive for this conversion than in the case of switching or
long-haul transmission.

20. Countries in which the Japanese have built optical fiber
systems include Argentina, Brazil, Taiwan, Hong Kong, and Saudi
Arabia. See, for example, "Japan's strategy for the 1980's: The
coming assault on communications markets," Business Week, 14
December 1981, and "Japanese now target communications gear
as a growth industry," The Wall Street Journal, 13 January 1983.

21. The smallest switches, ranging down to two lines, are
typically key telephone systems that take their name from the

switch or "key" used in early versions. Modern key systems have push buttons built into the telephone, each of which is connected to a central rack containing electronic circuitry that provides telephone functions, such as ringing, intercom, and call forwarding.

22. OECD, op. cit., pp. 46.

23. "IT&T: Groping for a new strategy," Business Week, 15 December 1980, as cited in OECD, op. cit., p. 46.

24. Switch manufacturers pursuing digital development programs include Western Electric, ITT, L. M. Ericsson, Siemens, GTE Automatic Electric, Nippon Electric Company, Fujitsu, Hitachi, Thomson-CSF, CIT-Alcatel, Northern Telecom, Plessey, and Stromberg-Carlson.

25. For a discussion of the competitive pressures surrounding export sales, see OECD, op. cit., pp. 46-47.

26. While intra-OECD exports of telecommunications equipment increased by a factor of 3.5 over the period 1970 to 1978, exports of this equipment from the OECD countries to the oil-exporting developing countries increased by a factor of 9 and to the other developing countries by a factor of 4.5. Telecommunications Equipment Industry Study, OECD, op. cit., p. 46.

27. For a discussion of OECD export opportunities in the developing countries, see OECD, op. cit., p. 47.

28. OECD, op. cit., p. 48.

29. The benefits of U.S. manufacture are also indicated by comparison of manufacturing labor costs. Though a relatively small portion of total costs, direct labor costs are currently lower in the United States than in Canada or western Europe.

30. The OECD has estimated that telecommunications equipment manufacturers account for 15 to 20 percent of world semiconductor production and for 10 to 15 percent of world semiconductor consumption. Recent acquisitions of semiconductor firm interests by telecommunications equipment manufacturers, particularly important for European firms that have been weak in the semiconductor field include: Advanced Micro Devices (Siemens), Litronix (Siemens), SEMI (GTE), and Signetics (Philips). OCED, op. cit., p. 49.

31. As in the computer industry, software has taken on increasing importance in recent years. For example, ROLM, one of the most highly successful PBX manufacturers, attributes its market position to its lead in developing special feature software aimed at solving customer problems such as route optimization to cut toll costs and controls to monitor long-distance calling. Some 50 percent of ROLM's development staff is currently working in software, double the percentage of seven years ago. See "A hot new challenger takes on Ma Bell," Business Week, 12 February 1979, and "Japan: Coming assault on the communications market," Business Week, 14 December 1981.

32. The 1968 Carterfone decision permitted customer equipment to be connected to the public telephone network. This was followed the next year by an authorization for protective interface devices, which allowed AT&T to charge customers for couplers supplied by the company to interface between data communications and telephone lines. In 1977 a registration program eliminated all need for protective devices.

Outside the United States, greater restrictions have been placed on competition. However several countries, including Japan, Canada, and the United Kingdom, are moving to ease these constraints.

33. U.S. Department of Commerce, U.S. Industrial Outlook 1982.

34. Gerald W. Brock, The Telecommunications Industry: The Dynamics of Market Structure, Harvard University Press, Cambridge, Massachusetts, 1981.

35. OECD, op. cit., p. 67.

7
The Consumer
Electronics Industry

U.S. firms were pioneers in consumer electronics technology and, until the 1960s, accounted for the largest share of world revenues and profits. In 1955 U.S. product shipments in consumer electronics were valued at $1.5 billion; Japanese firms produced only $70 million.[1] Nearly three decades later the situation is reversed, with Japanese revenues in consumer electronics more than twice those of U.S. manufacturers.

This chapter compares the performance of the U.S. consumer electronics industry with the competitive successes of the Japanese industry. The organization of the chapter is as follows. The first section describes the dimensions of the U.S. industry, its products, size, growth, and international position. The second section examines the dynamics of international competition in color television receivers, which represent the largest share of consumer electronics sales. Despite an early U.S. lead, U.S. firms today produce far fewer color television sets than do their Japanese counterparts. The initial penetration of Japanese firms in the United States was built on a foundation of labor-cost advantages, exploitation of mass merchandise distribution systems, specialization in small product sizes, and government protection. Ultimately, however, the Japanese established a worldwide dominant position based on aggressive technology, investment, and marketing strategies.

The third section of the chapter reviews the emergence of video cassette recorders (VCRs), a high-growth product used in conjunction with television sets. U.S. firms are conspicuously absent from the list of major developers and manufacturers of consumer video recorders. In contrast, several Japanese firms undertook sustained long-term development efforts that were ultimately rewarded with commercial success.

The final section evaluates the ingredients of the Japanese success in consumer electronics. The availability of a well-trained work force and a favorable environment for capital investment facilitated Japanese efforts to develop new products, such as the VCR, and to take the lead in advancing color television technology in key areas, such as the application of integrated circuitry and automated manufacturing. At the same time, however, industry protection was critical to the development of the Japanese color television industry, contributing to the ability of Japanese firms both to fund research and capital investment and to pursue aggressive pricing strategies that weakened the financial position of their U.S. competition.

INDUSTRY SIZE AND INTERNATIONAL POSITION

The output of the U.S. consumer electronics industry consists primarily of entertainment products--television receivers, video disk players, automobile radios, phonographs, radio-phonograph combinations, stereo compact systems, high-fidelity system components, autosound systems, loud speakers, microphones, and related products. During the last two decades, the composition and product orientation of the industry has changed substantially. The monochromatic and audio segments of the industry have grown slowly, while the color television segment has expanded rapidly. In 1981, for example, sales of color television receivers accounted for 58 percent of U.S. consumer electronics industry shipments.[2]

Conspicuous by their absence from the list of U.S. consumer electronics products are consumer radios, audio tape-recorder players, and video cassette recorders. Production of radios and audio tape recorders, which began to shift to Japan and elsewhere in the Far East in the 1950s, no longer exists in the United States. Volume production of VCRs originated in Japan, with some parallel development by Philips in Europe. The United States does not produce any VCRs at this time.

In 1981 total U.S. sales of consumer electronics products, supplied by domestic industry shipments and imports, were $9.8 billion.[3] Of this amount, imports--85 percent of which came from the Far East--stood at $5.9 billion. During the 1972 to 1980 period, U.S. imports rose at a 12.8 percent compound annual rate.

The majority of consumer electronics products imported into the United States are in finished form. These include television receivers, audio and video tape recorders, radios, high fidelity components, and loudspeakers. A significant portion of imports, however, are color receiver printed circuit boards with mounted components destined for assembly with cabinets and picture tubes in the United States. Japanese firms produce these boards and tuners in Japan for assembly in the United States, and U.S. manufacturers produce them in Mexico and the Far East for their U.S. plants.

Exports of consumer electronics products were $1.1 billion in 1981.[4] Exports--principally television receivers (assembled and unassembled), automobile radios, loudspeakers, amplifiers, and sound recorders and players--increased at a compound annual rate of 19.9 percent between 1972 and 1981. This rate, which was due primarily to the start of color broadcasting and adoption of the U.S. transmission standard in several Latin American countries, is not expected to be sustained in the face of increasing competitive pressure from the Far East.

Recent increases in U.S. consumer electronics shipments largely reflect the establishment of U.S.-based production facilities by foreign manufacturers following the negotiation of orderly marketing agreements (OMAs) with Japan in 1977 and with Taiwan and Korea in 1979.[5] In 1975, 3 out of 13 U.S. color television producers were foreign-owned; by 1980 foreign-owned firms accounted for 9 out of 15 firms in the U.S. industry. Of the nine foreign firms, seven are Japanese, one Taiwanese, and one Dutch.

Color Television

U.S. firms pioneered the development of color television in the 1940s and 1950s. RCA, the major contributor, began color television sales in 1954.[6] Color television, however, was slow to win widespread acceptance. Not only was color television a new mode of entertainment, but, at the time it was introduced, only a few programs per week were broadcast in color. Indeed RCA did not earn a profit on its color television sales until the early 1960s. By 1961 the only major color television manufacturers were RCA, Packard Bell, Magnavox, and Zenith, all of which had established reputations in monochrome television.

Color television, however, took off in the United States in the 1960s. Sales increased from $47 million in 1960 (of $797 million total TV receiver sales) to $2 billion in 1969, or 80 percent of all televisions sold.[7] Although the color television was a U.S. innovation, the spectacular growth in color television sales in the 1960s did not cement the position of U.S. firms as leaders in the world television manufacturing industry. Instead, by 1969, Japanese firms were well on the way to displacing U.S. firms as the dominant worldwide producers of video equipment.

Just as their own domestic demand began to grow rapidly in the mid-1960s, Japanese firms began marketing color televisions in the United States. Their initial penetration was made in private-label sales, where established U.S. brand names and distribution networks could be used and where low prices based on

Japanese labor-cost advantages were important. In 1964, for example, Toshiba signed a high-volume sales contract with Sears, Roebuck & Company. Japanese manufacturers also reportedly gave dealers higher profit margins as an incentive to promote Japanese products. As a result, Japanese firms achieved quick domination of the lowest cost distribution channels.

The Japanese private label penetration coincided with a shift in the structure of U.S. television distribution. Sales through mass-merchandise stores increased substantially in the mid- and late-1960s. This shift in distribution was facilitated by the increasing reliability of television sets, which reduced both the need for trained servicemen and the existing advantage of U.S. manufacturers with established service networks. As a result, private-label sales increased from approximately 13 percent of the U.S. total in 1966 to over 20 percent by 1970.

The early Japanese success in the United States was also aided by a void in the product line of U.S. firms in small-screen sets. As in other industries, Japanese firms specialized initially in smaller models, partly due to the nature of their domestic demand, which emphasized small, portable home furnishings. Moreover, U.S. firms in the face of a booming market for color television in the mid-1960s concentrated on larger, more profitable models.

RCA, the industry's technological leader in the 1950s and 1960s, shared its color television technology under license with firms around the world. RCA's willingness to license Japanese firms rather than to use its patents as a wedge to enter the Japanese market stands in sharp contrast to the strategies pursued by IBM in computers and Texas Instruments in semiconductors. Although this licensing policy generated significant income over the years for RCA, it also facilitated entry into the industry by Japanese firms.

Sony's introduction of its Trinitron system in 1968 greatly increased the picture quality of small-screen color sets. At the same time, Matsushita began placing increased emphasis on its Panasonic brand in the United States. The increase in the Japanese share of U.S. color television sales from the late 1960s through the mid-1970s largely corresponds to the success of the brand-name products of Sony and Matsushita, including the latter's acquisition of the Quasar brand from Motorola in 1974.

The Japanese gains in the 1970s were also based on their early application of integrated circuit technology and related process innovations.[8] Despite some exceptions, the Japanese television receiver industry as a whole appears to have adopted integrated circuitry faster than the U.S. industry. This solid-state design gave Japanese sets an advantage in reliability and compactness over the sets of the lagging U.S. firms. Related to their adoption of integrated circuitry, Japanese firms also moved aggressively to reduce the number of components and adopt automatic insertion techniques. These process innovations offset escalating Japanese labor and material costs and the appreciation of the yen that had by 1974 closed much of the gap between the price of Japanese and U.S. sets in the United States.

In 1975 the Japanese share of U.S. color television sales stood at 24 percent, with the Sony and Matsushita (Panasonic and Quasar) brands accounting for 14 percent. In contrast, in 1968, only 11 percent of U.S. color sales, largely under private label, were accounted for by Japanese imports. By 1976 much of the remaining share of private-label production was taken over by Japanese firms, increasing the Japanese share of U.S. sales to approximately 33 percent.

Three of the U.S. brands that declined in the face of Japanese competition in the 1970s--Magnavox, Philco, and Sylvania--have been acquired by Philips, the dominant European producer. That gives Philips the third largest U.S. market share, after RCA and Zenith. RCA and Zenith maintained their U.S. market shares of approximately 20 percent each during the 1970s, though their profitability declined. The only other U.S. brand name producer is General Electric, which has increased its U.S. market share in recent years.[9]

Mounting concern in the United States over the rise in Japanese imports culminated in the negotiation of an OMA with Japan in 1977. This agreement resulted in a 25 percent drop in imports from Japan in that year and a continuing drop through 1980. In 1979 additional OMAs were concluded with Taiwan and Korea.[10] In 1981, when the OMA with Japan lapsed, imports of complete receivers increased an estimated 52 percent, an increase that is largely attributable to the combination of the overvalued U.S. dollar and the undervalued Japanese yen.

Although Japanese imports dropped between 1977 and 1980, the OMA with Japan yielded little change in Japanese market share. Instead the OMA resulted in a change in the location of production as Japanese manufacturers acquired or built production facilities in the United States. Indeed, the number of U.S.-owned firms declined from 18 in 1968 to 5 in 1981, while the number of foreign-owned companies increased from none to 9.[11] U.S. plants with Far Eastern ownership now supply 30 percent of U.S. industry shipments of color TV receivers.

Japanese penetration of European sales occurred later than in the United States, but has reached similar proportions. (The European color television industry, unable to agree on a technical standard until 1966, did not grow significantly until the early 1970s.)[12] In 1971, 95 percent of Japanese color exports went to the United States; by 1974 nearly 30 percent went to Europe. Japanese gains in Europe have increased further in recent years, particularly since the expiration of patents on the PAL color broadcast system, which was the system adopted in West Germany and the United Kingdom. PAL patent licenses included restrictions limiting the size of imported sets to 19 inches or less, which is smaller than the most popular European sets. Japanese firms have also set up production and assembly facilities in Europe, as they did in the United States, and entered into joint ventures with European firms.

Consumer Video Cassette Recorders

Video recording, like the transistor and color television, was a U.S. invention. The first practical video tape recorder bringing important changes in the broadcasting of TV programs was introduced by Ampex Corporation in 1956.[13] The Ampex machine, called the Quadruplex, generated worldwide sales and set the standard for broadcasting use for two decades. Although RCA began producing video tape machines in 1959, Ampex continued to dominate sales to broadcasters.

The Quadruplex machine was a massive, complex, and expensive machine filling a large console and two equipment racks and selling for $50,000 in its monochrome version. The complexity was necessary to produce a signal that met the stringent requirements of broadcast use. Experiments with an alternative approach, later termed helical recording, however, led to developments that promised recorders that were much simpler to make and use than the Quad machines. Early helicals produced pictures that would look quite adequate in quality for the general public, but which were inadequate for broadcasters. Although it was clear that the helical design might be suitable for a host of uses outside of broadcasting, less clear was which of many paths to follow in developing the technology, how to develop new markets for the resulting products, and how good a business it would be once products and markets were developed.

The home video cassette recorder was developed step by step over 20 years, interactively by nearly a dozen companies worldwide. During the 1960s, firms in the United States, Japan, and Europe participated in the technical and commercial development of helical recording. Outside Japan the leaders were Philips, which dominated European professional and broadcasting sales of video recorders, and Ampex, which extended its broadcast leadership in the United States with a line of professional and industrial units. However, neither Philips nor Ampex was focusing on a consumer product at this time. None of the leading U.S. consumer electronics firms invested significantly in video recording until after 1970.

In Japan, on the contrary, eight or more companies--including all leading consumer electronics manufacturers--launched aggressive efforts to develop helical video recording technology. Sony and Matsushita, among the first to succeed in marketing a consumer product, held the goal of achieving a design suitable for the consumer market from the very beginning--even though they sold their first-generation products to other types of users.

By 1970 the first-generation helical cassette machines-- developed by RCA and Cartravision in the United States, Philips in Europe, and by several companies in Japan--were ready for demonstration. Despite optimistic predictions that the age of cartridge television in the home had arrived, it took an additional five years to develop and market the first successful video cassette recorder. The only commercially successful products at this stage--offered by Philips and Sony--were destined for professional and industrial use.

Then, in 1975, Sony launched the now-legendary Betamax. Within two years, Japan Victor, adopting some of Sony's innovations and adding some variations of its own, perfected an alternative design. Termed VHS (Home Video System), it was adopted by Japan Victor's parent firm, Matsushita, and now shares the bulk of world sales with Sony's Beta format. Matsushita announced the production of their two millionth VHS machine in late 1980; Sony's sales of the Betamax reached 750,000 units in 1980 alone. The sole competitor to these product formats is an innovative Philips design manufactured also by Grundig and sold in Europe.

Sony, Japan Victor, and Matsushita made many important innovative contributions to the development of the VCR. Nevertheless, their Betamax and VHS machines also contain many elements invented by firms such as Ampex, Philips, and Toshiba, whose success in the VCR field has been more limited. Substantial and sustained development efforts over a lengthy period with a consumer product as the ultimate goal, rather than superior inventive performance, thus appears to have been the critical ingredient in the success of the three current market leaders. Moreover, each of these firms maintained a strategic commitment that kept development going, even when prematurely commercialized consumer products failed. At the same time, they were investing in the development of advanced manufacturing techniques and the expansion of production capacity.

THE INGREDIENTS OF THE JAPANESE SUCCESS IN CONSUMER ELECTRONICS

U.S. manufacturers, in the view of many observers, made miscalculations in corporate strategy and management practice, not only in the case of color television and video cassette recorders, but also in the earlier cases of monochrome TVs and transistor radios.[14] Although U.S. firms such as Zenith and RCA were responsible for most of the major advances in color television technology throughout the 1960s and 1970s, the Japanese focus on manufacturing issues such as higher productivity, improved quality, and reliability enhanced their cost competitiveness. Even when domestic demand was brisk, they built positions in export markets, beginning with the largest--the United States-- and aiming at market segments overlooked by the U.S. industry. After initial success, they broadened their product lines and deepened penetration.

The ability of Japanese consumer electronics firms to undertake long-term commitments to innovation strategies with uncertain payoffs has been facilitated by the three major factors--the availability of a well-trained work force, low capital costs, and industry protection--that provide the potential for Japanese success in other segments of the electronics industry. First, as discussed in Chapter 2, Japanese firms have been able to draw on an educated and stable work force, including a high proportion of electrical engineering graduates. For example, a U.S. Department of Commerce survey found that Japanese consumer electronics firms in the early 1970s employed roughly twice the R&D manpower as U.S. firms, despite R&D spending that was at approximately the U.S. level.[15] In the context of lifetime employment, this human resource base has served as an important asset for Japanese innovators.

Second, favorable capital costs and availability have supported Japanese consumer electronics companies in the pursuit of greater quality and efficiency in the exploitation of export markets and in the development of novel products that promise long-term rather than short-term results. Japanese firms made commitments to consumer applications of video recording 15 years before the demand actually could be tapped. They persisted in their efforts to develop the basic technology, even when prematurely introduced consumer products failed. Moreover, in both video recording and color television, they invested significantly in advanced manufacturing processes that offered high quality and productivity.

Finally, Japanese firms received important benefits from their location in Japan. The Japanese consumer electronics industry was protected until Japanese firms were well established in terms of technology and production scale. By 1970, for example, production volume of color television in Japan exceeded the U.S. level. Yet investment was restricted, and the Japanese tariff remained above the U.S. level. In addition the Japanese market was protected during the 1960s and early 1970s by a myriad of non-tariff barriers, including the distribution system, the commodity tax, and the availability of foreign exchange.[16]

In all of their consumer electronics product lines, Japanese firms thus served a large, protected domestic market that provided cash flow and revenue growth. Moreover, the Japanese market was not fragmented; the leading firms had large shares, giving them a significant scale of operations. This large, concentrated, and protected domestic market allowed Japanese firms to pursue a two-tier pricing strategy in Japan and the United States--a factor that precipitated the color television dumping controversy. Low U.S. prices weakened the ability of U.S. manufacturers to fund R&D and capital investment; high Japanese prices strengthened the financial position of their Japanese competitors. Thus while U.S. manufacturers also possessed a large and concentrated market, they lacked two things the Japanese had from the start--access to an even larger foreign market using the same technical standards (the United States) and protection against import competition. Perhaps more important, they lacked an aggressive public policy mechanism to invoke against discriminatory pricing.

NOTES

1. Richard S. Rosenbloom and William J. Abernathy, "The Climate for Innovation in Industry: The Role of Management Attitudes and Practices in Consumer Electronics," Research Policy, forthcoming.

2. U.S. Department of Commerce, U.S. Industrial Outlook 1982.

3. Ibid.

4. Ibid. U.S. exports go mainly to Canada and Latin America. Recent export growth has been concentrated primarily in Latin America.

5. The increased internationalization of the U.S. color television industry is also reflected in related-party transactions as a percentage of total transactions. (These transactions are defined as a measure of the flow of material between parent firms and foreign subsidiaries.) Between 1975 and 1979, related-party transaction figures for incomplete receivers and subassemblies from Japan, Taiwan, Singapore, and Mexico ranged from 68.7 percent to 99.8 percent. Japan showed the largest increase--from 68.7 percent in 1975 to 91.3 percent in 1979, reflecting the shift in the later stages of the Japanese manufacturers' production process to the United States. U.S. Department of Commerce, U.S. Industrial Outlook 1982.

6. This description of the evolution of the color television industry draws on the following sources: Donald G. Fink, "Perspectives on Television: The Role Played by the Two NTSC's in Preparing Television Services for the American Public," Proceedings of the IEEE, September 1976; Michael Radnor, et al., The U.S. Consumer Electronics Industry and Foreign Competition, Northwestern University Center for the Interdisciplinary Study of Science and Technology, Evanston, Illinois, May 1980; Charles River Associates, International Technological Competitiveness: Television Receivers and Semiconductors, prepared for the National Science Foundation, 1979; Merton J. Peck and Robert W. Wilson, "Innovation, Imitation, and Comparative Advantage: The Performance of Japanese Color Television Set Producers in the U.S. Market." in Herbert Giersch, ed., Emerging Technologies: Consequences for Economic Growth, Structural Change and Employment," J. C. B. Mohr, Tubingen, West Germany, 1982.

7. Electronics Industries Association, Electronic Industries Year Book, Washington, D.C., various years.

8. The Japanese industry's move to ICs was aided by the Japanese government's decision to reduce the high commodity tax on solid-state color television sets and by a cooperative R&D effort to apply ICs to television receivers that began in 1966 and involved researchers from industry, the universities, and government laboratories.

9. "TV: A growth industry again," Business Week, 23 February 1981.

10. The share of imports of color television receivers from countries other than Japan increased from less than 2 percent in 1976 to 9 percent in 1980, reflecting increased offshore sourcing by U.S. and Japanese firms as well as some increases by firms based in Korea and Taiwan.

11. U.S. Department of Commerce, U.S. Industrial Outlook.

12. As a result of the European delay in adopting technical standards for color broadcasting, European demand was suppressed until well after the period normally required for technology to flow abroad. As a result, the U.S. industry's early innovative lead in color television did not give U.S. firms an advantage in Europe.

13. The description of the development of the VCR in this section is based on Rosenbloom and Abernathy, op. cit.

14. See, for example, J. C. Abegglen and W. V. Rapp, "The Competitive Impact of Japanese Growth," in Jerome B. Cohen, Pacific Partnership: United States--Japan. Japan Society, Inc., Lexington, Massachusetts, 1972; U.S. General Accounting Office, "Color Television," United States-Japan Trade: Issues and Problems, September 1979; "Japan's strategy for the 80s," Business Week, 14 December 1981.

15. U.S. Department of Commerce, The U.S. Consumer Electronics Industry, U.S. Government Printing Office, Washington, D.C., 1975.

16. For a discussion of these trade barriers as they relate to the case of color television, see U.S. General Accounting Office, "Color Television," United States-Japan Trade: Issues and Problems, September 1979.

8
Policy Options for the U.S. Electronics Industry

The U.S. electronics industry's record of growth and innovation over the past three decades has been remarkable; continued success in the 1980s, however, cannot be taken for granted. Internationally, the world has entered a new era of fierce competition in which U.S. firms find themselves competing against foreign governments that have targeted their domestic industries to surpass U.S. technology leaders. Domestically, the overall U.S. economy faces considerable uncertainty. At the same time, pressure on state and federal budgets will make more difficult the investment in education and research that is needed to strengthen the industry's infrastructure.

These domestic and international challenges demand coordinated policy responses by industry, government, and universities. They call for a new emphasis on specific industrial policies that will encourage the flow of investment to new and innovative technologies. These include (a) policies to ensure adequate research and development by U.S. firms; (b) investment policies that foster a steady flow of capital for new ventures as well as for expansion and modernization; (c) education policies designed to develop the engineers, computer scientists, technicians, and technologically aware citizens needed for today's society; and (d) international trade and monetary policies that promote fair access to world markets.

This chapter outlines the building blocks of an industrial strategy designed to encourage technological innovation and investment in electronics. The following four sections address the range of research, capital formation, human resource, and international trade and monetary policy options from which a comprehensive U.S. competition policy for the electronics industry may be forged. The final section provides a postscript to the report.

104

RESEARCH POLICY

A strong technological position is vital if the U.S. electronics industry is to maintain its long-run competitiveness in the face of a persisting Japanese cost-of-capital advantage. Essential to the U.S. industry's continued technological leadership is a high level of basic or fundamental research. Such research involves long time horizons, uncertain commercial outcomes, and the possibility of low appropriability and, therefore, places substantial disincentives before all but the very largest firms.[1] As discussed in Chapter 2, however, the ability of even the largest firms--and the universities--to maintain their historical commitment to long-term fundamental research is threatened by both high capital costs and shortages of engineers and computer professionals at the doctoral level.

Technological innovation also requires a commitment to applied research and product development if basic scientific advances are to be translated into marketable products. Here, too, the ability of the electronics industry to undertake sustained investment is hindered by high capital costs and shortages of well-trained engineers and computer professionals. Indeed, applied R&D--especially development--consumes the major share of industrial R&D expenditures, while innovation-related capital investment requires an even higher level of spending beyond the R&D stage. It is at these investment-intensive stages that the Japanese electronics industry, with its cost-of-capital advantage and its strong supply of engineers, poses the greatest challenge to U.S. firms.

The technological leadership of the U.S. electronics industry in commercially oriented R&D also faces a challenge based on the growing willingness of foreign national governments to finance research in electronics. Pressures toward increasing this involvement are coming from several directions. First, R&D costs, particularly in the area of software development, have increased across all segments of the electronics industry as integrated circuit design and manufacturing techniques have advanced. Further, intensified international competition has made it more difficult for some firms to finance long-term commercially oriented R&D. At the same time, increased government interest in electronics reflects its growing national security and defense importance, its future role in terms of output and employment, and its productivity-enhancing impact on other industries.

The governments of France, West Germany, the United Kingdom, and Japan have all funded major commercially oriented electronics research programs aimed at enhancing the international competitiveness of their national industries. Such direct funding overcomes the disincentives faced by private firms in financing research with long gestation periods, namely, uncertainty with respect to commercial outcomes and appropriability. Moreover, as demonstrated by the Japanese, cooperation by private firms in joint research efforts does not mean that such firms will not be fiercely competitive in subsequent development, production, and marketing efforts.

In contrast to the national commerical R&D policies of other major industrialized countries, U.S. government electronics research funding is overwhelmingly oriented to its national defense and aerospace programs.[2] Although commercial development has not been an objective of these programs, they have nevertheless exerted a major continuing influence on the competitiveness of U.S. electronics products ranging from computers, lasers, fiber optics, radio and television equipment, to robotics and communication satellite technology.

Large-scale defense and aerospace contracts have provided U.S. electronics firms with a ready demand, for which they have expanded production and thereby gained valuable experience, know-how, and scale economies. Moreover, the willingness of the U.S. Department of Defense (DOD) to pay a premium for quality and reliability has helped electronics firms bear the cost of refining and debugging their products. Due in part to the demand of the DOD and the National Aeronautics & Space Administration (NASA) for faster and more reliable semiconductors, the U.S. semiconductor industry was able to reduce its unit costs quickly during the 1960s, solidifying its position as a world technology leader.

Other influences associated with defense and aerospace funding have been less beneficial. The successful marketing of new products requires long lead times during which firms can apply new technologies and make sure they have adequate capital, labor, and productive capacity to meet anticipated demand. Defense and NASA programs, however, are subject to relatively sudden changes in national security needs and prevailing politics. Between 1967 and 1974, in the wake of Vietnam, defense-related R&D declined by $3.7 billion (in constant 1972 dollars), drastically reducing the nation's demand for scientists and engineers. In contrast, the rise in defense spending since 1981 threatens to create bottlenecks in the production of key subcomponents and capital goods and shortages of engineers and scientists in advanced electronics.

Finally, commercialization requires that new technology be transferable to commercial uses at relatively low cost. Since the 1960s, however, military technological requirements have been increasingly divergent from commercial applications. As a result, many observers expect that commercial spillovers from DOD- and NASA-funded research will be relatively small.[3]

Although the U.S. government has provided little direct funding for long-term research with commercial objectives, substantial indirect support has historically been provided under the guise of federal policy toward the structure of the telecommunications industry. Bell Laboratories, for example, has received steady, long-term funding for its basic research budget under license contract fees paid by the Bell operating companies and, therefore, indirectly by most of the U.S. population in the form of telephone service and equipment charges. Bell scientists have been able to pursue research interests with great latitude and publish research results freely. This favorable working environment has, in turn, attracted the top scientists that have made Bell Labs one of the preeminent research laboratories in the world.

The modus operandi of Bell Labs, however, will change as a result of the 1982 Consent Decree between AT&T and the U.S. Department of Justice (DOJ).[4] The divestiture of AT&T's local service operating companies will eliminate a major source of Bell Lab's funding for basic research. This change could force Bell Labs to reduce the magnitude of its basic research or the latitude enjoyed by its researchers. Such changes would diminish Bell Lab's ability to attract top scientists and to push technology across as many fields and at the same rate that it has done in the past.[5] Moreover, under the recent consent decree, AT&T will no longer be required to follow an open-licensing policy.[6]

The next years will tell whether the corporate tradition of technology that exists in AT&T will sustain Bell Labs' commitment to basic research. Other cooperative approaches to basic research, however, are likely to increase, due to a growing number of advocates in industry, government, and the universities. In the past, U.S. firms have avoided cooperative research due in part to the effect of antitrust laws, despite the fact that there appears to be little reason to expect that cooperation between participants in the earliest stages of research will diminish competition in their subsequent product development efforts. Although the DOJ has the power to grant approval to joint research ventures that make them immune to federal antitrust prosecution, such immunity does not extend to treble-damage antitrust action by competitors against the participants. Recent moves to modify the antitrust laws offer a more equitable approach, immunizing an applicant from any retroactive prosecution from the time the DOJ approves a specific venture until the project is completed or the Department considers it injurious to the competitive balance for the venture to continue and, thus, revokes any future immunity.

Major firms in both the semiconductor and the computer industries have recently proposed electronics research ventures as a means of better utilizing R&D dollars and manpower. One such

venture, the Microelectronics and Computer Technology Corporation, has been formed by a group of electronics firms led by
Control Data Corporation to conduct joint long-range programs,
e.g., in microelectronics packaging, advanced computer architecture, computer-aided design and manufacture, and software productivity.[7] Participants will either be shareholders, who will
finance one or more technology programs, or associates with more
limited involvement.

Another joint research venture, the Semiconductor Research
Cooperative, has been established by the Semiconductor Industry
Association as an affiliate nonprofit research cooperative that
will fund basic research at U.S. universities.[8] Contributions by
participants are to be assessed at one-tenth of 1 percent of their
annual sales of semiconductors, with no one company bearing
more than 10 percent of the total annual budget. Although such
joint funding reduces the disincentives associated with basic
research, particularly for smaller firms, it does not eliminate
them since research findings are expected to be published and,
thus, freely available. Nevertheless, participants in the
cooperative are likely to have a lead-time advantage over
competitors in the commercialization of technological advances.

Finally, tax legislation passed in 1981 provided increased
financial incentives for commercial R&D.[9] The key provision,
which began in 1981 and runs through 1985, was a 25 percent tax
credit for certain incremental R&D expenditures (above the
average outlay for the previous three years) in the United States.
(Salaries of support staff and the cost of nonsalary benefits for
researchers do not qualify for the tax credit.) Tax benefits were
also extended to companies donating certain kinds of equipment
to universities for research. Many industry observers, however,
believe that these changes do not provide a major stimulus to
R&D, though they may bias the system in the right direction.
Additional proposals thus call for expanding the incremental R&D
tax credit to include every category of R&D expenditures as well
as research grants to universities for projects related to a firm's
business. As discussed below, such incentives offer the advantage
of directly encouraging greater industry-university cooperation in
long-term research.

CAPITAL FORMATION POLICY

The U.S. electronics industry's need for capital has grown
dramatically in recent years. The industry must invest in new
production capacity, not only to meet demand for advanced
products, but also to upgrade existing facilities rendered obsolete
by the rapid change in process technology. At the same time, the

capital intensity of production has grown substantially, first in the manufacture of integrated circuits and, more recently, in computers and computer-related equipment.

Failure to provide a satisfactory climate for long-run risk taking in an industry that is as research intensive as electronics will severely penalize the future competitiveness of the industry. In the face of record high interest rates, however, U.S. electonics firms have experienced increasing difficulty in raising capital to invest in the R&D and production capacity necessary to keep up with rapidly changing technology. In contrast, Japanese firms face a highly favorable domestic environment for investment in product innovation and production capacity. Due to both a high domestic savings rate and a tightly controlled capital market, interest rates in Japan have been generally lower than those in the United States and western Europe.[10] Low interest rates, in conjunction with the high debt-to-equity ratios permitted by the Japanese financial system, have given Japanese firms a cost-of-capital advantage that has allowed them to follow more aggressive capital investment strategies than their international competitors.

Low capital costs give Japanese electronics firms an edge in price-sensitive standardized products where high-volume manufacturing skills are essential. The ability of Japanese firms to invest in automated production capacity, often well in advance of anticipated demand, has been a critical ingredient of the Japanese success in products ranging from color television receivers to low-priced computer peripheral equipment and strategically important MOS computer memories. By establishing themselves in such high-volume products, the Japanese have, in turn, moved into a position from which they can challenge the U.S. technological lead in lower volume, large system products across a broad spectrum of the electronics industry.

In the face of the Japanese challenge, policies to create a more favorable climate for capital formation in the United States, such as tax incentives to stimulate savings and investment, have become the focus of increasing attention. The full impact of such policies goes beyond the scope of this report to the reindustrialization of the nation; nevertheless, one policy--depreciation reform--goes to the heart of the unique capital formation needs of the electronics industry. In the semiconductor industry, for example, the equipment used to manufacture integrated circuits becomes technologically obsolete long before the end of its physical life. Process innovations are so frequent that, on average, existing process techniques have been superseded by new innovations every two years. Yet the U.S. tax code--despite the reforms of 1981--continues to discriminate against industries where the economic life of equipment can be substantially less than the permissible tax-based life, which is based on physical rather than technological obsolescence.

Reforms of the depreciation system in the United States thus would require recognition of the rapid technological obsolescence in the electronics industry. One such reform could involve the creation of a category for equipment with a high rate of technological obsolescence, combined with a reduction in the penalty for taking the investment tax credit over a short time period. A second reform, an increase in the first-year depreciation allowance, would provide a substantial incentive for new capital investment.

General policies to encourage growth and innovation, such as depreciation reform, offer an advantage in that they do not require the government to pass judgment on the merits of a particular investment. Any tax or depreciation schedule in support of new investment has the effect of raising the overall level of technology, but leaves the judgment of the usefulness of specific technologies to the entrepreneur and the marketplace. Indeed many observers stress that general tax policies rather than affirmative actions in support of particular technologies have been the cornerstone of Japan's highly successful industrial policy.

Finally, policies that affect the availability of venture capital and equity capital investment, such as the tax treatment of capital gains, employee stock options, and corporate dividends, can play an important role in the electronics industry given its high research requirements, growth, and risk relative to other industries.[11] Such policies are particularly important to smaller firms that do not have easy access to capital markets, especially when they are seeking to enter the industry on the basis of relatively high-risk innovations. In 1969, for example, the introduction of more restrictive tax treatment of capital gains contributed to a dramatic reduction of entry by new firms in the semiconductor industry. Only following revision of the capital gains tax law in 1978 did the number of new ventures again begin to climb.

HUMAN RESOURCE POLICY

The response of the U.S. electronics industry to the opportunities and challenges of the decades ahead will depend critically on its ability to attract sufficient numbers of engineers, scientists, and technicians of the highest possible quality. Since World War II the growing complexity of electronics technology has generated an ever increasing demand for engineers and computer professionals. Yet in the critical field of electrical engineering, the United States produced only 14 percent more bachelor's degree graduates in 1980 than it did in 1970, while the number of master's degree graduates decreased by 12 percent, and the number of doctorates awarded decreased by 40 percent. Trends for computer professionals have been similar; namely, rising undergraduate enrollments and declining Ph.D. production. These trends are distressing, particularly in comparison with Japan, which has long been graduating more engineers per capita than the United States and now graduates more in absolute terms.

Although market forces have worked to ease shortages at the bachelor's level, they have simultaneously exacerbated the shortage of qualified faculty in engineering schools and computer science departments. As electronics firms have expanded their research efforts, they have lured faculty members away from academic research into well-paid positions in industry. At the same time, industry is making such attractive job offers to bachelor's degree recipients that many who would once have gone to graduate school now opt for positions in industry.

Several factors in addition to noncompetitive salaries contribute to the problem of attracting and retaining qualified faculty members: difficulties in obtaining research support, problems of inadequate equipment and facilities, and the instability of government funding for research and fellowships. Further, the current shortage of graduate students and faculty members creates unusually heavy teaching loads, which makes academic jobs less attractive for those interested in research. The net effect has been a reduction in the ability of universities to provide education in engineering and the computer professions, although undergraduate demand for these areas is more intense than ever. Unless the problem of faculty erosion is alleviated, many engineering schools and departments that educate computer professionals may be forced to reduce their enrollments during this decade.

Compounding the seriousness of the shortage of qualified faculty is a severe lack of equipment. At the undergraduate level, for example, equipment required to teach computer-aided design and manufacturing methods is generally unavailable in engineering schools. Consequently, a good deal of the bachelor's instruction being offered may in fact be obsolete. At the graduate level, the increasing cost and relatively short useful life of experimental facilities, particularly in the integrated circuit area, pose substantial barriers to the educational process.

The strained capacities of engineering and computer science departments reflect the changing requirements of industry. As such, they call for closer industry-university cooperation in anticipating and preparing for future demands. Universities must recognize the special research needs of their engineering and computer science faculties. One alternative is to adapt engineering schools to the so-called medical school model, whereby faculty members would be allowed more liberty to supplement their salaries and gain access to specialized research facilities in industry.[12] At the same time, universities can allow faculty members increased opportunities to undertake research projects with the sponsorship and participation of industry.[13]

Industry, provided with appropriate incentives by government, can take important steps to respond to the shortage of faculty and equipment. Steps discussed above include the formation of consortia to support university research groups and the provision of money, equipment, and personnel in exchange for university-conducted research. Given adequate capacity and safeguards, companies can make their unique research facilities available to university faculties. Firms can offer cooperative arrangements so that university faculty members can engage in industrial research, while industrial engineers serve in university departments.

In the past two years, the electronics industry has taken significant steps to increase its support for engineering education. In 1981 the American Electronics Association recommended that electronics firms give 2 percent of their annual R&D expenditures to help universities hire more engineering faculty and purchase computer equipment. Since then, a growing number of firms—Control Data, General Electric, Hewlett-Packard, IBM, Xerox, Wang—have introduced fellowship programs, equipment grants, and other means of helping universities train electrical and computer engineers.

Although greater corporate support can make an important contribution to alleviating the shortage of qualified faculty and obsolescent equipment, the government must continue to be the major contributor to the process. Federal agencies, such as the National Science Foundation (NSF), DOD, and NASA, can assist the universities in areas ranging from the purchase of equipment to the development of incentives to encourage students in graduate engineering and computer programs to enter university teaching.[14] State governments can increase their support for university electronics programs, and a growing number have done so.

Restoring engineering and science education as a national priority will require a new partnership among federal, state, and local governments, and private interests. One model is a "High-Technology Morrill Act" that would authorize matching grants for nonfederal initiatives, not just from state governments but also from high-technology industries.[15] Industrial grants, for example, could stimulate grants in accordance with a designated formula from both state and federal governments. Such an act would create a long-term funding mechanism for engineering and science education, one in which industry and universities together would play a major role in determining where investments in education should be made.

Even with a concentrated effort by industry, universities, and government, it will take three to five years before the size of the existing engineering work force can be increased with new graduates. Therefore, in the near term, industry and academic leaders have called for new efforts to increase the productivity of the pool of experienced engineering talent. The pace of technological change has been so rapid that practicing engineers are finding it more and more difficult to stay abreast of the latest technical developments. Too many engineers leave the profession each year for better paying and more influential jobs in management. This trend must be reversed and a solution found so that more talented people stay in the field to cope with increasingly demanding technical problems. One approach that deserves careful examination is the recent Massachusetts Institute of Technology proposal for a nationwide council to promote lifelong education for working engineers.[16]

Beyond policies to support professional science and engineering education is the need for a program designed to strengthen the scientific and technological education of all U.S. citizens. The role of science and technology is increasing throughout society in business, in government, in the military, and in occupations and professions where it never before intruded. Today people in a wide range of nonscientific and engineering occupations and professions must have a greater understanding of technology than at any time in history. Yet, over the past 15 years the U.S. educational system has placed declining emphasis on science and mathematics--in contrast to other industrialized countries. More high school students than ever before are dropping out of science and mathematics courses after the tenth grade, and this trend shows no signs of abating. Unless reversed, the current trend toward virtual scientific and technological illiteracy will not only undermine the competitive strength of the U.S. electronics industry, but the social fabric of the nation as a whole.

A renewal of a national commitment to excellence and international primacy in science, mathematics, and technology will require a coordinated program to (a) increase public awareness of the need for excellence in science and technology and (b) help the schools fulfill their role in formal science and technology education. Initiatives in the first area must be directed at society as a whole and, thus, will depend on the cooperative and complementary activities of many sectors, including business and industry, the federal government, state and local governments, community organizations, and university and industrial scientists and engineers. In the second area, specific programs must focus on schools, helping them do a better job of producing graduates who are prepared to function in an increasingly technology-oriented society. To this end, educational experts have proposed major initiatives aimed at improving curricula, introducing new electronic technologies, helping teachers, and increasing awareness of career opportunities in science and technology. Although such measures are not a panacea, they can go a long way toward strengthening the nation's science and engineering capabilities.

INTERNATIONAL TRADE POLICY

Foreign markets account for over half of the total value of sales in the electronics industries analyzed in this report. This fact alone marks the importance of international trade for U.S. firms. The changing economics of the electronics industry, however, have heightened the importance of international trade, not only to U.S. firms, but also to electronics firms worldwide. Maximum access to world markets is essential to keep production costs as low as possible, particularly in high-volume products where learning curves and scale economies may be substantial. Similarly, the larger the potential market, the more easily escalating research and development costs can be recovered .

Japan and other major producing countries have responded to this situation by placing greater weight on implications for international competitiveness when framing national policies. For example, public subsidies for long-term research in microelectronics are seen in many countries as a means of improving the international competitiveness of domestic suppliers. At the same time, however, competitive pressures have intensified as the number of countries seeking to maintain or build a domestic capability in electronics has increased. National governments have thus sought in some cases to further the cause of their domestic industries by restricting access of foreign competitors through both tariff and non-tariff barriers. In Japan, foreign firms face a formidable number of non-tariff barriers, not the least of which is Japan's propensity not to import. Impediments to successful direct investment in Japan by foreign firms range from difficulties in recruiting able and experienced engineers to preferential access for Japanese companies to capital, government guarantees, special tax incentives, loans, and subsidies.[17] In the European Economic Community (EEC), firms from nonmember countries face punitive tariffs (though the EEC selectively suspends duties on products for which production is inadequate or nonexistent in member states), highly restrictive rules of origin that limit the value of imported components in a finished product to less than 5 percent, and discriminatory government procurement policies in individual member states.

Other trade-distorting actions include subsidization of exports, countertrade requirements that require a supplier to take goods that it would not otherwise buy, and the application of political pressure on potential purchasers.[18] The United States, for example, provides tax subsidies to the foreign subsidiaries of U.S. firms through domestic international sales corporations as well as low-interest rate financing to purchasers through the U.S. Export-Import Bank. Such measures can transform international competition between firms into competition between countries. A case in point is provided by the telecommunications equipment industry, where OECD countries have traditionally been closed to foreign producers, and exports to non-OECD countries rest on a relatively small number of very large contracts.

The preference for domestic production will, in the long run, be self-defeating if restricted market access distorts or hinders the role of competition in diffusing the benefits of technological innovation.[19] Most countries are too small on their own to support broad-line domestic manufacturers in major segments of the electronics industry. However, in an industry as large as electronics, there is considerable room for intra-industry specialization without industry protection. Indeed, the potential for specialization has increased with the rapid advance of electronics technology.

To be effective, trade liberalization must involve some agreement as to the rules of the game for international competition in the electronics industry, as well as the adoption of a better mechanism for settling bilateral disputes within GATT. In many countries the electronics industry has benefited from direct and indirect financial support. Public involvement in electronics will continue in the years ahead as governments seek to promote legitimate social and economic policy objectives, e.g., by organizing, financing, and carrying out basic research in electronics. Such support, however, will adversely affect trading relations between countries if it distorts or hinders international competition. For example, the refusal of the Japanese government to allow U.S. subsidiaries or U.S.-Japanese joint ventures to participate in its VLSI project caused considerable resentment in the U.S. semiconductor industry. To avoid resentments and frictions, agreement is therefore needed as to what constitutes appropriate subsidies and countervailing duties.

Rather than sliding into protectionism, the United States must organize so that it can expand trade, keep it fair, and keep itself competitive. Eight U.S. cabinet departments currently have statutory roles in international trade policy: State, Treasury, Agriculture, Defense, Commerce, Labor, Transportation, and Energy. In addition, there are five important independent agencies involved in trade policy plus the Office of the U.S. Trade Representative within the Office of the President. The time has come to clarify and consolidate responsibility for foreign trade policy within one agency. Otherwise, divided authority will increasingly confuse both domestic and foreign trade interests, thwarting an aggressive U.S. posture on unfair pricing and other non-tariff barriers.

The trading efforts of U.S. firms have also been inhibited by restrictive laws and regulations, such as the Foreign Corrupt Practices Act, numerous export controls, and confusing antiboycott legislation, despite the fact that trade policy has not worked well as an instrument of foreign policy. Indeed, the spector of U.S. trade sanctions has increased the local sense of vulnerability in Japan and western Europe, adding fuel to protectionist sentiment. In the long run, sanctions may do costly damage to the U.S. reputation as a dependable supplier among customers all over the world. Review and clarification of restrictive U.S. laws and regulations are essential to eliminate instances where they place U.S. firms at a competitive disadvantage.[20]

Finally, progress toward freer world trade must address the underlying causes of the repeated, and severe, exchange-rate misalignments that have periodically emerged between the dollar and the yen.[21] Exchange-rate misalignments go far in explaining the recent escalation in international trade tensions. Over the past three years, the yen has become substantially undervalued--dramatically improving Japan's price competitiveness in the world economy. At the same time, the dollar has become substantially overvalued--undermining U.S. competitiveness both in Japan and elsewhere. From its lows in late 1978 to its highs of August 1981 and April 1982, the dollar rose by 35 to 40 percent against the yen. Meanwhile Japanese inflation ran about 20 percentage points less than U.S. inflation. The price competitiveness of the United States in world trade vis à vis Japan thus deteriorated by 50 percent or more within three years.

The result has been disastrous for U.S. exports. The strength of the U.S. dollar has in effect created an export barrier for U.S. products. Part of the problem is the excessively high interest rates that have made dollar investments enormously attractive, sustaining the dollar at an artificially high level. The role of the U.S. Federal Reserve System in determining the value of world currencies thus merits careful examination. The impact of its decisions is no longer merely domestic but worldwide. At the same time, the world monetary system--especially the dollar's role as the world's principal reserve currency--requires study and reform if massive exchange-rate misalignments are to be avoided in the future. Unless the dollar's value comes down, U.S. electronics exports will be severely handicapped, regardless of any success in eliminating other barriers to trade.

SHAPING AN ELECTRONICS INDUSTRY POLICY

A program to enhance the competitiveness of the U.S. electronics industry requires major policy initiatives in research and development, capital formation, education and training, international trade, and monetary reform. The costs and benefits of policies in each area must be debated in an open public forum in order to achieve a broad-based consensus involving the industry, government, and universities. Such agreement is necessary if policies to promote the industry are to be formulated and administered with an understanding of their combined effects.

Nor can a competitive policy for the electronics industry be seen in a vacuum independent of the economy as a whole. Upgrading and improving the competitiveness of other industries will provide an important stimulus for the electronics industry. Indeed electronics-based technologies, e.g., CAD/CAM and robotics, are frequently the key to improving productivity in other industries. Modernization and rationalization in other industries will increase the demand for both electronic systems and components, leading to lower costs and prices and further increases in demand.

Since this NAE panel study was initiated in 1980, notable steps have been taken to move to a constructive dialogue about the long-term strategic issues confronting the U.S. electronics industry. The industry has proposed major reforms to help it adjust and compete in increasingly competitive international markets. The efforts of firms in the computer and semiconductor industries to promote cooperative research are indicators of the growing support for the development of a national program to achieve long-term technological objectives. Industry-university cooperation has increased, and new corrective proposals have been advanced, particularly by leading engineering educators. Both the federal and state governments have introduced tax reforms aimed at invigorating the climate for innovation by electronics firms.

Much remains to be done. Capital costs for the increasingly capital-intensive electronics industry must be controlled. A greater share of the U.S. GNP should be devoted to savings, investment, and innovation. A national policy for private investment that would permit more rapid capital recovery and encourage entrepreneurial investment is much needed, as are efforts to eliminate disincentives to saving. The U.S. government needs to develop and convey long-term policy intentions to the public to build confidence for saving and investment.

Future shortages of technical personnel are likely to hinder the ability of the electronics industry to innovate and, hence, maintain its competitiveness. The U.S. educational system has faltered as the number of new engineering doctorates has declined. The number and quality of technical graduates must increase to keep pace with the growing numbers of engineers and computer professionals in Japan and other competitor nations.

Trade barriers to U.S. exports into competitor nations must be reduced. The United States is viewed by its competitors as the most attractive of world electronics markets and has been targeted in the national plans of Japan and France for their direct attack. The United States must pursue a more effective program to promote fair trade among international competitors or face the inevitable protectionist pressure to increase U.S. barriers to imports. Finally, U.S. trade policy must be coordinated with its international monetary policy.

The U.S. government, working with industry and the universities, must develop a statement of national goals for the electronics industry. This nation has traditionally had an aversion to any effort associated with national planning, particularly if it is to be done by government. Yet other nations are, within their systems, clearly planning and marshalling major resources to enhance the long-term international competitiveness of their industries. The published Japanese plans, articulated by MITI, are but one example. The challenge to the United States is to formulate an industrial policy that will foster the ability of its electronics firms to compete vigorously in world markets.

NOTES

1. On the question of appropriability, for example, industry observers are currently debating whether the nation's patent and copyright laws should be modified to afford increased protection to innovations in semiconductor circuitry and computer microcode.

2. The federal government accounts for approximately 30 percent of all U.S. R&D funding, two-thirds of which is devoted to defense and space. The missions of DOD and NASA, however, are strongly developmental in nature. In 1980, for example, basic research support accounted for 5 percent of the DOD's research funding and for 9 percent of NASA's. Although other federal government agencies, e.g., the National Science Foundation and the National Bureau of Standards, fund electronics research, these sources are small in magnitude compared to defense and space. National Science Board, Science Indicators 1980.

For a discussion of the influences of the research funding of DOD and NASA, see Robert B. Reich, "Making Industrial Policy," Foreign Affairs, Spring 1982, and Nestor E. Terleckyj, "Analyzing Economic Effects of Government R&D Programs," National Planning Association working paper, 8 December 1981.

3. Another factor limiting the commercialization of both defense and other government-funded research has been restrictions with respect to the licensing of resulting patents. Changes in federal patent policy in 1980, however, have made it possible to obtain exclusive licenses for development achieved while working under federal contract. This change may encourage industrial research that might have been foregone in the past when its fruits would pass immediately into the hands of competitors.

4. See, for example, "The organizational switches at Bell Labs," The Wall Street Journal, 19 July 1982, and "Bell Labs: The threatened star of U.S. research," Business Week, 5 July 1982.

5. The top researchers lost from Bell Labs would presumably pursue their research activities at other institutions. However, it is debatable whether the same number of people working together in particular areas with the freedom from funding pressures and from other constraints (e.g., academic teaching loads and academic committee responsibilities) that have traditionally characterized Bell Labs could be assembled in other settings.

6. Under the 1956 antitrust consent decree, AT&T was required to license Bell Labs' patents to other companies on a nondiscriminatory basis. Such open licensing has facilitated entry by new firms across all segments of the electronics industry. Although AT&T will no longer have to license Bell Labs patents under the terms of the new consent decree, its licensing policy is likely to remain relatively open for several reasons. Cross-licensing is highly institutionalized in electronics due in part to the systems nature of products (with interdependent patents held by different firms) and in part to the ease with which other firms can invent around a particular patent. In addition, other large patent holders such as IBM follow open-licensing policies. Finally, overly restrictive licensing could provoke antitrust challenges. Indeed antitrust actions initiated during the 1940s and 1950s contributed to the widespread licensing that currently prevails in the electronics industry.

7. After a start-up period, the annual budget of the Microelectronics and Computer Technology Corp. is expected to be $50 to $100 million. "U.S. electronics firms form venture to stem challenge by Japanese," The Wall Street Journal, 26 August 1982.

8. The Semiconductor Research Cooperative expects to spend $6 million in the first year and $10 to $15 million in its second year. While the universities would hold all patents and issue licenses, a participant's financial contributions would be considered as prepayment of royalties. "Semiconductor makers expand research fund," The New York Times, 14 April 1982.

9. Tax incentives, of course, are only effective if corporations are subject to a significant corporate tax. To the extent that effective U.S. tax rates are more moderate than comparable

Japanese tax rates, tax incentives will be a less potent policy instrument in the United States than in Japan.

10. Although Japan has recently moved to ease its restrictive capital market policies, its capital markets remain the most tightly controlled of any major economy. See, for example, "Are the Japanese rigging the yen?" Fortune, 31 May 1982; "Is Japan holding the yen down?" Business Week, 8 March 1982; "Borrowing yen will be a little bit easier," Business Week, 31 May 1982; and "Borrowers are eager to get yen loans but must grapple with Japan's delays," The Wall Street Journal, 7 July 1982.

11. Such tax policy changes have typically been made in response to the strength of lobbying efforts by interested parties. In 1976, for example, Congress eliminated favorable tax treatment of employee stock options--a significant part of the compensation that small, growing firms use to attract and hold talented employees--in the absence of effective lobbying effort by growth companies. See, for example, "Lobbyist say options tax break is needed to spur innovation, and Congress responds," The Wall Street Journal, 1 July 1981.

12. There is considerable debate about whether and to what extent engineering schools should adopt this organizational model and divorce themselves from their traditional association with faculties of arts and sciences. This debate focuses on the appropriate content of the engineering curriculum, particularly the optimum balance between science-based training and training in engineering design, development, and production.

13. A major catalyst in the spectacular expansion of high-technology industries along Route 128 around the rim of Boston and in Silicon Valley south of San Francisco was the presence of major research universities willing to allow faculty to work with private business.

14. The National Science Foundation is especially crucial to science and engineering education. Recently, the NSF's education budget has been subject to sharp cutbacks, declining from $70 million in fiscal year 1981 to $20 million in 1982 and $15 million in 1983. The proposed budget for 1984, however, would reverse this trend.

15. For a discussion of this and other policy proposals, see, James Botkin, Dan Dimancescu, and Ray Stata, "High Technology, Higher Education, and High Anxiety," Technology Review, October 1982, and U.S. Department of Education and National Science Foundation, Science and Engineering Education for the 1980s and Beyond, October 1980.

16. The MIT proposal recommends a council that would consist of participants drawn from the universities, private companies, research organizations, and professional societies and would be run by a board of directors composed of chief executives

of member organizations. As proposed, its basic mission would be to present a range of approaches by acting as a clearinghouse for information about educational needs, resources, and existing industry-university cooperative programs. Beyond that, it is proposed that the council (1) take an active role in securing more industry and university commitments to new educational programs tailored to the needs of working engineers, (2) raise and distribute funds for new programs, (3) identify and enlist services of engineers and university faculty members in developing new courses of study, and (4) organize, monitor, and evaluate pilot teaching of newly developed courses. See James D. Bruce, William M. Siebert, Louis D. Smullin, and Robert M. Fano, Lifelong Cooperative Education, MIT Department of Electrical Engineering and Computer Science, October 1982.

17. In May 1982 Japan announced a package of trade reforms designed to liberalize its domestic market in response to mounting pressure from the United States and the European Economic Community. Since then, however, the Japanese have made little progress in implementing these measures. The resulting U.S. frustrations over trade have, in turn, spawned highly protectionist proposals in Congress to require reciprocity and higher U.S.-made content in imported products. See, for example, "Japan: U.S. reprisals loom as trade reforms stall," Business Week, 23 August 1982. In January 1983, reacting to mounting criticism, Japan announced still another package of measures designed to dismantle its non-tariff barriers to trade by simplifying product standards and certification procedures. Again, however, the legislation necessary to implement such reforms may take months or years. "Japanese adopt another package to open market," The Wall Street Journal, 14 January 1982.

18. The role of countertrade has increased in recent years, spurred by government efforts to maintain exports to recession-shrunken markets and to narrow balance-of-payments deficits. For U.S. policymakers, the proliferation of such practices poses a dilemma: whether to try to halt the spread of countertrade or help U.S. firms to get their share. See, for example, "New restrictions on world trade," Business Week, 19 July 1982.

19. Protectionist measures that restrict access to some markets also limit the innovation incentives and funding ability of private firms located outside those markets. Firms based in restricted markets face both their own protected domestic demand plus demand in the open market. To the extent that protected firms face a larger potential sales base than outside firms, they will possess a greater incentive to innovate. In addition a large protected sales base enhances the ability of firms located in such markets to fund R&D, to achieve scale and learning curve economies, and to engage in international price

discrimination. Price discrimination, or "dumping," reduces the ability of firms in the more open markets to fund R&D because lower prices reduce their profitability.

20. One promising development, however, is the 1982 Export Trading Company Act, which reduces two hurdles that posed impediments to the formation of export trading companies in the United States. (Under the new legislation, an export trading company is defined as any group of companies and banks that joins forces with the specific objective of selling goods and services abroad.) First, the new law removes a major deterrent to joint export ventures by offering the possibility of prior certification of an antitrust exemption. This antitrust exemption will allow traders to take advantage of economies of scale; one firm, for example, might represent a number of manufacturers selling the same product abroad--handling marketing, packaging, and warehousing; arranging transportation and insurance; preparing documents for customs; distributing goods; and servicing foreign customers on behalf of the sellers after the sales are made. Second, the measure also amends the longstanding prohibition barring banks from investing in commercial enterprises. Bank-holding companies, with their considerable financial strength, may for the first time take a direct equity interest in export trading companies, thereby increasing the access of export trading companies to export services and financing. The new law will thus serve to reduce the costs and risks of international trade, particularly for small and medium-sized firms that would otherwise be inhibited from exporting by financial constraints and lack of familiarity with foreign markets, customs, and laws.

Although export trading companies existed in the United States prior to the new legislation, those firms have typically been thinly capitalized ventures specializing in one or two export services such as insurance, ocean shipping, finance, licensing, or market research and development. Two of the leading U.S. exporters are actually Japanese trading companies, Mitsui and Mitsubishi, which are licensed to sell U.S. goods as part of their worldwide activities.

21. For a detailed discussion of foreign exchange rate policy problems, see C. Fred Bergsten, "What To Do about the U.S.-Japan Economic Conflict," Foreign Affairs, Summer 1982.

Biographical Sketches

FERNANDO JOSE CORBATO is Professor of Computer Science and Engineering at the Massachusetts Institute of Technology. He holds a B.S. and a Ph.D. from the California Institute of Technology and the Massachusetts Institute of Technology, respectively. He has been with the Massachusetts Institute of Technology since 1956. Dr. Corbato is a Fellow of the Institute of Electrical and Electronics Engineers and received their W. W. McDowell Award in 1966. He is a member of several honorary and professional organizations, including the National Academy of Engineering, and was co-author of The Compatible Time Sharing System and Advanced Computer Programming.

THERESE FLAHERTY is an Assistant Professor at the Harvard Business School. She received a B.A. from Tufts University and a Ph.D. from Carnegie Mellon University. Before joining the Harvard faculty, Dr. Flaherty was Assistant Professor of Economics at Stanford University. Her special areas of interest are in economics, industrial organization, and management of technology in global companies. She has published several articles related to international competition in the semiconductor industry.

EUGENE I. GORDON is Director of the Lightwave Devices Laboratory at Bell Laboratories. He took a B.S. degree from City College of New York and a Ph.D. degree from the Massachusetts Institute of Technology. He joined Bell Laboratories in 1957. Dr. Gordon serves as chairman of the Working Group on Imaging and Display Devices of the Advisory Group on Electron Devices for the Office of the Director of Defense Research and Engineering. He is a member of a number of professional and honorary organizations, including the National Academy of Engineering. He is a Fellow of the Institute of Electrical and Electronics Engineers and received their 1975 Valdimir M. Ivorykin Award.

Dr. Gordon is the author of 43 published articles and holds 23 patents.

WILLIAM C. HITTINGER is Executive Vice-President of Research and Engineering at the RCA Corporation. He holds a B.S. from Lehigh University, did graduate studies at Stevens Institute, and was awarded an honorary doctorate degree in engineering from Lehigh University. Before joining RCA in 1970, Mr. Hittinger served as President of Bellcom, Incorporated and of General Instrument Corporation. He is a member of several professional and honorary organizations, including the National Academy of Engineering, is a Fellow of the Institute of Electrical and Electronics Engineers, and is the author of many publications.

ANNETTE LAMOND is a consultant specializing in industrial organization economics. She holds a B.A. degree in economics from Wellesley College, an S.M. in management from the Massachusetts Institute of Technology, and a Ph.D. in economics from Yale University. Dr. LaMond is a member of Phi Beta Kappa and is the author of several books and articles, including a paper on the competitive status of the U.S. semiconductor industry to be published by the Harvard Business School.

JOSEPH C. R. LICKLIDER is Professor of Electrical Engineering and Computer Science at the Massachusetts Institute of Technology. He received A.B. and A.M. degrees from Washington University and a Ph.D. from the University of Rochester. Dr. Licklider was a Vice-President of Bolt Beranek & Newman, Incorporated, and Director of Behavioral Science and Information Research for the Advanced Research Project Agency of the U.S. Department of Defense. He is a member of the National Academy of Sciences and a fellow of several professional organizations. In 1954 Dr. Licklider received the Franklin V. Taylor Award from the Society of Engineering Psychologists.

JOHN G. LINVILL is Professor of Integrated Systems and Director of Industrial Programs at the Center for Integrated Systems at Stanford University. He holds an A.B. from William Jewell College and S.B., S.M., and Sc.D. degrees from the Massachusetts Institute of Technology. His professional experience was with the Massachusetts Institute of Technology and Bell Telephone Laboratories before joining Stanford University in 1955. He is an active participant in many professional and honorary organizations, including the National Academy of Engineering. He was awarded an honorary doctorate degree in applied sciences by the Catholic University of Louvain, Belgium, and has received two awards for development of the

Opticon, a reading aid for the blind, as well as the John H. McAulay Award from the American Association of Workers for the Blind and the Education Medal from the Institute of Electrical and Electronics Engineers. Dr. Linvill holds nine patents, has authored or co-authored seven books, and has published numerous articles and technical reports.

ROBERT N. NOYCE is Vice-Chairman of the Board of Intel Corporation. He received a B.A. at Grinnell College and a Ph.D. at Massachusetts Institute of Technology. He has worked with Philco Corporation and Schockley Semiconductor Laboratory. Dr. Noyce was one of the founders of Fairchild Semiconductor, where, as Research Director, he was responsible for the initial development of the Silicon Mesa and Planar transistor lines. When the company became a division of Fairchild Camera and Instrument Corporation, he was elected a Vice-President of the Corporation. Dr. Noyce was awarded the Stuart Ballantine medal from the Franklin Institute and is a Fellow of the Institute of Electrical and Electronics Engineers, a member of the American Physical Society, and a member of the National Academy of Engineering. He holds 15 patents.

DANIEL I. OKIMOTO is Assistant Professor of Political Science at Stanford University. He has studied at the University of Tokyo and holds a B.A. from Princeton University, an M.A. from Harvard University, and a Ph.D. from the University of Michigan. Before assuming his present position, Dr. Okimoto was affiliated with The RAND Corporation and Seikei University. He was a Mellon Foundation Fellow at the Aspen Institute for Humanistic Studies and a National Fellow at Hoover Institution. He is a member of various professional organizations and is author or co-author of numerous publications.

M. KENNETH OSHMAN is President and Chief Executive Officer of ROLM Corporation. He received B.A. and B.S. degrees from Rice University and M.S. and Ph.D. degrees from Stanford University. His area of specialization is electrical engineering. Before founding ROLM Corporation, Dr. Oshman was employed by GTE Products, Inc. Dr. Oshman was the recipient of the Entrepreneur of the Year Award, Peninsula Chapter of the Stanford University Business School Alumni Association in 1977; the Outstanding Engineer Award, Texas Society of Professional Engineers in 1964; and the Engineering Alumni Award, Rice University in 1963. He is a member of the National Academy of Engineering.

MICHAEL RADNOR is Director for the Center for the Interdisciplinary Study of Science and Technology and Professor in the J. K. Kellogg Graduate School of Management at Northwestern University. He holds a B.S. degree and a Diploma from the London School of Economics, a B.Sc. degree and a Diploma from the Imperial College of Science and Technology at London University, and a Ph.D. degree from Northwestern University. Prior to joining the faculty at Northwestern University, his professional experience included senior manufacturing engineer for Westinghouse Electric and Vice-President and General Manager of Tann Controls Company. Dr. Radnor served as principal investigator and principal author of a study of the U.S. consumer electronics industry for the U.S. Department of Commerce. He is author or co-author of numerous books and many articles. Dr. Radnor holds membership in a number of professional organizations.

WILLIAM V. RAPP is Commercial Officer with the International Trade Administration of the U.S. Department of Commerce. He took a B.A. at Amherst College, an M.A. at Stanford University, and an M.A and Ph.D. at Yale University. Dr. Rapp has served as a Vice-President with the Bank of America and with the Morgan Guaranty Company. His expertise is in investment banking.

ROGER S. SEYMOUR is currently a consultant after retiring as Program Director for Corporate Staff Commercial Relations, IBM Corporation. Mr. Seymour holds a B.S. in industrial administration from Yale University.

ROBERT W. WILSON is a consultant specializing in industrial organization economics. He received an S.B. degree in physics from the Massachusetts Institute of Technology and a Ph.D. in economics from Yale University. He is the author of several articles and books concerning competition, regulation, and technological change. Prior to establishing a consulting practice, Dr. Wilson was an economist with the Antitrust Division of the U.S. Department of Justice.

This report on the electronics industry is one of seven industry-specific studies (listed below) that were conducted by the Committee on Technology and International Economic and Trade Issues. Each study provides a brief history of the industry, assesses the dynamic changes that have been occurring or are anticipated, and offers a series of policy options and scenarioes to describe alterntive futures of the industry.

The Competitive Status of the U.S. Auto Industry,
ISBN 0-309-03289-X; 1982, 203 pages, $13.95

The Competitive Status of the U.S. Machine Tool Industry,
ISBN 0-309-03394-2; 1983, 78 pages, $5.95

The Competitive Status of the U.S. Fibers, Textiles, and Apparel
Complex, ISBN 0-309-03395-0; 1983, 90 pages, $7.95

The Competitive Status of the U.S. Pharmaceutical Industry,
ISBN 0-309-03396-9; 1983, 102 pages, $8.95

The Competitive Status of the U.S. Ferrous Metals Industry,
ISBN 0-309-03398-5; approx. 135 pages, $9.95 (prepublication
price), available Spring 1984

The Competitive Status of the U.S. Civil Aviation Manufacturing
Industry, ISBN 0-309-03399-3; approx. 120 pages, $9.25
(prepublication price), available Spring 1984

Also of interest . . .

International Competition in Advanced Technology: Decisions for
America ". . .should help mobilize Government support for the
nation's slipping technological and international trade position. . . .
Leonard Silk, the New York Times. A blue-ribbon panel created by
the National Academy of Sciences takes a critical look at the
state of U.S. leadership in technological innovation and trade.
ISBN 0-309-03379-9, 1983, 69 pages, $9.50

Technology, Trade, and the U.S. Economy, ISBN 0-309-02761-6,
1978, 169 pages, $9.75

Quantity discounts are available; please inquire for prices.

All orders and inquiries should be addressed to:

Sales Department
National Academy Press
2101 Constitution Avenue, NW
Washington, DC 20418